ACS SYMPOSIUM SERIES 302

Coulombic Interactions in Macromolecular Systems

Adi Eisenberg, EDITOR
McGill University

Fred E. Bailey, EDITOR
Union Carbide Technical Center

Developed from a symposium sponsored by
the Macromolecular Secretariat
at the 188th Meeting
of the American Chemical Society,
Philadelphia, Pennsylvania,
August 26–31, 1984

American Chemical Society, Washington, DC 1986

Library of Congress Cataloging-in-Publication Data

Coulombic interactions in macromolecular systems.
(ACS symposium series, ISSN 0097-6156; 302)

Bibliography: p.
Includes indexes.

1. Macromolecules—Congresses. 2. Ionomers—Congresses.

I. Eisenberg, A. (Adi) II. Bailey, Fred E., 1927–
III. American Chemical Society. Macromolecular Secretariat. IV. American Chemical Society. Meeting (188th: 1984: Philadelphia, Pa.)

QD380.C68 1986 547.7′0457 86-3641
ISBN 0-8412-0960-X

PRINTED IN THE UNITED STATES OF AMERICA

ACS Symposium Series

M. Joan Comstock, *Series Editor*

FOREWORD

The ACS SYMPOSIUM SERIES was founded in 1974 to provide a medium for publishing symposia quickly in book form. The format of the Series parallels that of the continuing ADVANCES IN CHEMISTRY SERIES except that, in order to save time, the papers are not typeset but are reproduced as they are submitted by the authors in camera-ready form. Papers are reviewed under the supervision of the Editors with the assistance of the Series Advisory Board and are selected to maintain the integrity of the symposia; however, verbatim reproductions of previously published papers are not accepted. Both reviews and reports of research are acceptable, because symposia may embrace both types of presentation.

CONTENTS

PREFACE

THE GENERAL IMPORTANCE of the effects of coulombic interactions on the properties of macromolecules is seen in the wide range of phenomena that are relevant to macromolecular science and engineering. This volume contains the majority of the papers presented at the Macromolecular Secretariat symposium held in Philadelphia in 1984. The sponsorship of this symposium by five ACS divisions, as well as the frequency of other symposia on this topic, both in the United States and in Europe, confirms the high level of activity in this area.

As in other Macromolecular Secretariat symposia, each participating division selected a topic it considered very active at the time the meeting was organized. In addition, two sessions dealt with more general aspects of coulombic interactions. Thus, the coverage of the topic is not encyclopedic, or even uniform, but is dictated solely by divisional judgments and areas of high activity.

It is a pleasure to express our gratitude to the divisional organizers, K. A. Mauritz (Polymeric Materials Science and Engineering), R. D. Lundberg (Rubber), W. J. MacKnight (Polymer Chemistry), D. A. Brant (Cellulose, Paper and Textile Chemistry), and C. Thies (Colloid and Surface Chemistry). Special thanks go to Rajagopalan Murali for his most valuable assistance in the editing of this volume.

ADI EISENBERG
McGill University
Montreal, PQ, H3A 2K6, Canada

FRED E. BAILEY
Union Carbide Technical Center
South Charleston, WV 25303

January 22, 1986

INTRODUCTION

1

Structure and Applications of Ion-Containing Polymers

R. A. Weiss[1], W. J. MacKnight[2], R. D. Lundberg[3], K. A. Mauritz[4], C. Thies[5], and D. A. Brant[6]

[1]Institute of Materials Science, University of Connecticut, Storrs, CT 06268
[2]Department of Polymer Science and Engineering, University of Massachusetts, Amherst, MA 01003
[3]Exxon Chemical Company, Linden, NJ 07036
[4]Department of Polymer Science, University of Southern Mississippi, Hattiesburg, MS 39406–0076
[5]Department of Chemical Engineering, Washington University, St. Louis, MO 63130
[6]Department of Chemistry, University of California, Irvine, CA 92717

The interest in the subject of ion-containing polymers has continued unabated since the development of organic ion-exchange resins in the 1940's. This interest is due to the variety of properties and applications that result from the interactions of ions bound to organic macromolecules. These interactions affect the physical properties as well as the transport properties of the host material, and these polymers have found applications in areas as diverse as thermoplastic elastomers, permselective membranes, and microencapsulation membranes. The common thread in all these applications and the ion-containing polymers that are used is that the ionic moieties and their interactions dominate the behavior of the polymer. It is not surprising, therefore, that the great bulk of the scientific effort that has been devoted to ion-containing polymers for the past two decades has been directed at delineating and understanding the very complex nature of these ionic interactions and the microstructure of the polymers. This effort has spawned a number of scientific conferences devoted to ion-containing polymers, monographs(1-5), and review articles(6-9) in addition to hundreds of scientific papers and industrial patents.

The purpose of this paper is to provide an overview of the field of ion-containing polymers. This review is not intended to be exhaustive but is meant to provide a suitable introduction to the field as well as demonstrate the diversity of applications of ion-containing polymers. The main emphasis will be on ionomers-- that is, polymers composed of a hydrocarbon or fluorocarbon backbone containing a small amount of pendent acid groups (usually less than 10 mol%) that are neutralized partially or completely to form salts. This sub-field of ion-containing polymers is emphasized because it is the one that is currently experiencing the greatest activity, both from a scientific

0097–6156/86/0302–0002$06.00/0

standpoint of understanding the structure-property relationships and from the technological development of new materials and applications.

This is not to say that more highly ionized polymers, such as conventional polyelectrolytes, are not of technological importance and interest. In fact, just the opposite is true. Polyelectrolytes have historically been utilized as ion-exchange resins, but a number of novel applications such as cements, gels, and encapsulation membranes are under development. Several applications of these materials, such as polyelectrolyte complexes and ionic biopolymers, are also included in this review.

STRUCTURE OF IONOMERS

Theory

The first attempt to deduce the spatial arrangement of salt groups in ionomers was that of Eisenberg (10), in which he assumed that the fundamental structural entity is the contact ion-pair. On the basis of steric considerations, he showed that only a small number of ion-pairs, termed "multiplets", can associate without the presence of intervening hydrocarbon and that there is a tendency for multiplets to further associate into "clusters" that include hydrocarbon material. This association is favored by electro-static interactions between multiplets and opposed by forces arising from the elastic nature of the polymer chains. Eisenberg assumed that the chains on average would undergo no dimensional changes as a result of clustering of the ionic species. Forsman (11) later removed this restriction and showed that the chain dimensions must actually increase as a result of association, a result confirmed by neutron scattering experiments (12).

There is a considerable body of experimental and theoretical evidence that salt groups in ionomers exist in two different environments, i.e., multiplets and clusters. The multiplets are considered to consist of small numbers of ion dipoles, perhaps up to 6 or 8, associated together to form higher multipoles -- quadrapoles, hexapoles, octapoles, etc. These multiplets are dispersed in the hydrocarbon matrix and are not phase separated from it. This means that in addition to acting as ionic crosslinks, they affect the properties of the marix, such as the glass transition temperature, water sensitivity, etc. Clusters are considered to be small (< 5 nm) microphase separated regions rich in ion pairs but also containing considerable hydrocarbon. They possess at least some of the properties of a separate phase and have a minimal effect on the properties of the hydrocarbon matrix, though they may have some reinforcing effect.

The proportion of salt groups that resides in either of the two environments in a particular ionomer is determined by the nature of the backbone, the total concentration of salt groups, and their chemical nature. Despite considerable research by various groups the details of the local structure of these materials remains somewhat obscure, as does the question of how low molecular weight polar impurities such as water affect the local structure.

Experimental Studies

1. X-ray Scattering

Small angle x-ray scattering (SAXS) results have been of central importance in the interpretation of the structure of ionomers. Figure 1 compares the x-ray scattering observed for low-density polyethylene, ethylene-methacrylic acid copolymer, and its sodium salt over a range of scattering angles (2θ) from 2° to 40°. Polyethylene crystallinity is present in all three samples, though the acid copolymer and the ionomer exhibit less crystallinity than polyethylene. The ionomer scattering pattern contains a new feature, a peak centered at about 2θ = 4°. This "ionic peak" appears to be a common feature of all ionomers, regardless of the nature of the backbone or cation and regardless of the presence or absence of backbone crystallinity.

Both the magnitude and the location of the ionic peak are dependent on the nature of the cation. Thus the ionic peak occurs at lower angles for cesium cations of a given concentration than for corresponding lithium cations. In addition, the magnitude of the ionic peak is much greater for cesium than for lithium. The ionic peak persists at elevated temperatures but disappears when the ionomer is saturated with water. The scattering profile, however, in the vicinity of the ionic peak in the water-saturated ionomer is different from that of the parent acid copolymer.

A molecular interpretation of scattering data is model dependent, and several models for the distribution of salt groups in ionomers have been proposed to explain the ionic peak. They consist mainly of two approaches: (1) that the peak arises from structure within the scattering entity, i.e., from intraparticle interference, and (2) that the peak arises from interparticle interference.

The "shell-core" model(13) originally proposed in 1974 and later modified(14,15) is representative of the intraparticle interference models. It postulates that in the dry state a cluster of ca. 0.1 nm in radius is shielded from surrounding matrix ions not incorporated into clusters by a shell of hydrocarbon chains, Figure 2. The surrounding matrix ions that cannot approach the cluster more closely than the outside of the hydrocarbon shell are attracted to the cluster by dipole-dipole interactions. This mechanism establishes a preferred distance between the cluster and the matrix ions; a distance of the order of 2 nm accounts for the spacing of the SAXS ionic peak.

Yarusso and Cooper(16) proposed a different interpretaiton of the ionic peak that considers the liquid like scattering from hard spheres described originally by Fournet(17). With this interparticle interference model, Yarusso and Cooper were able to quantitatively model the ionic peak for lightly sulfonated polystyrene ionomers. They found that for the zinc salts about half of the ionic groups are aggregated in well-ordered domains, i.e., clusters, with the remainder dispersed in the matrix as multiplets. The clusters are about 2.0 nm in diameter and approach each other no more closely than 3.4 nm center to center. Although based on quite different physical principles,

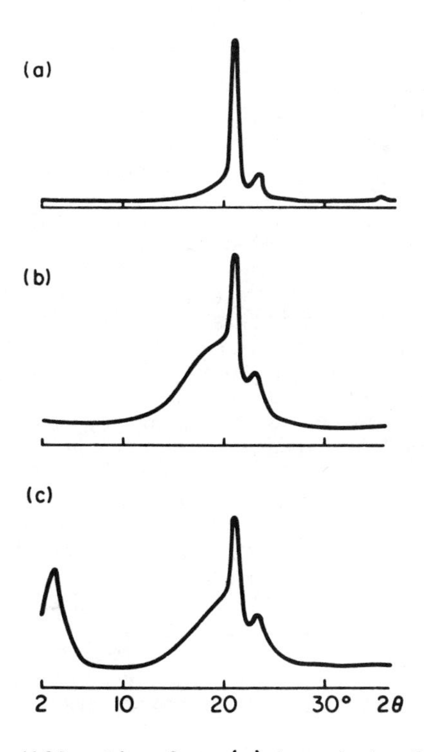

Fig. 1. X-ray diffraction from (a) branched polyethylene, (b) copolymer of 94% ethylene and 6% methacrylic acid, and (c) sodium salt of copolymer in (b), 90% neutralized. (reprinted with permission from ref 64)

this model yields structural parameters very similar to those obtained from the shell-core model.

2. Neutron Scattering

Small angle neutron scattering (SANS) has assumed great importance in the investigation of polymer morphology. One of its most impressive accomplishments is the measurement of single chain dimensions in bulk. This is generally achieved by selectively labeling a small percentage of the polymer chains by replacing hydrogen with dueterium in order to take advantage of the much higher coherent neutron scattering cross section of the deuteron compared to the proton.

Several SANS studies of ionomers have appeared on both deuterium labeled and unlabeled systems(8,12,18-20). The earlier work(14) showed that an ionic peak, similar to that observed by x-rays, could be discerned in some cases, especially when the sample was "decorated" by the incorporation of D_2O. It was also tentatively concluded(19) that the radius of gyration, R_g, of the individual chains is not altered when the acid is converted to the salt in the case of poly-styrene-methacrylic acid copolymers. Subsequent SANS experiments were performed on sulfonated polystyrene ionomers with up to 8.5% sul-fonation(12). The results of this study indicated that aggregation of the ionic groups is accompanied by considerable chain expansion, which is consistent with the theory of Forsman(11).

In a separate investigation, a series of polypentenamer sulfonate ionomers was studied(20). In this case, contrast was achieved by adding measured amounts of D_2O to the samples. Figure 3 shows the results for a 17 mol% cesium sulfonate ionomer. For the dry film there is no evidence of a scattering maximum. As small amonts of D_2O are added, however, the SANS peak becomes detectable. The Bragg spacing of the SAXS ionic peak observed for the same sample in the dry state is essentially the same as the SANS peak at low D_2O concen-trations. Above a D_2O/SO_3 ratio of about 6, the SANS ionic peak moves markedly to lower angles. These results are consistent with a phase separated model where absorbed water is incorporated into the ionic clusters, remaining separate from the matrix even at saturation.

3. Electron Microscopy

Reference 7 reviews a number of electron microscopy studies of ionomer morphology in the period up to 1979. None of these studies makes a convincing case for the direct imaging of ionic clusters. This is because of the small size of the clusters (less than 5 nm based on scattering studies) and difficulties encountered in sample prepara-tion. The entire problem was reexamined in 1980(21). In this study ionomers based on ethylene-methacrylic acid copolymers, sulfonated polypentenamer, sulfonated polystyrene and sulfonated ethylene-propylene-diene rubber (EPDM) were examined. The transfer theory of imaging was used to interpret the results. Solvent casting was found to produce no useful information about ionic clusters, and microtomed sections showed no distinct domain structure even in ionomers neutra-lized with cesium. Microtomed sections of sulfonated EPDM, however,

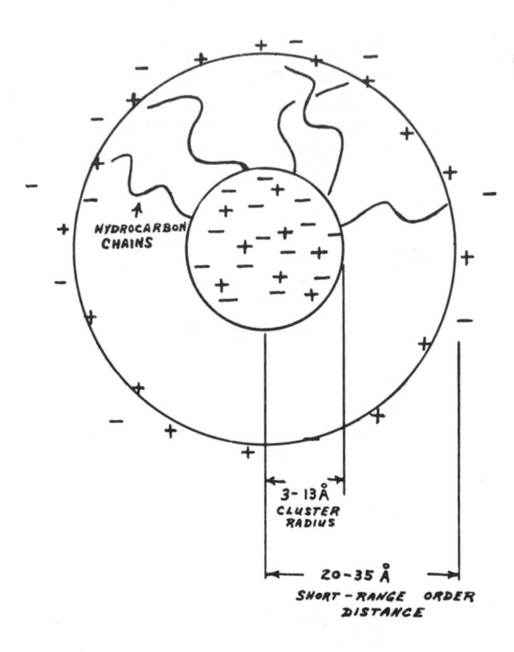

Fig. 2. Model of the ionomer sructure. (reprinted with permission from ref 65. Copyright 1974 John Wiley and Sons).

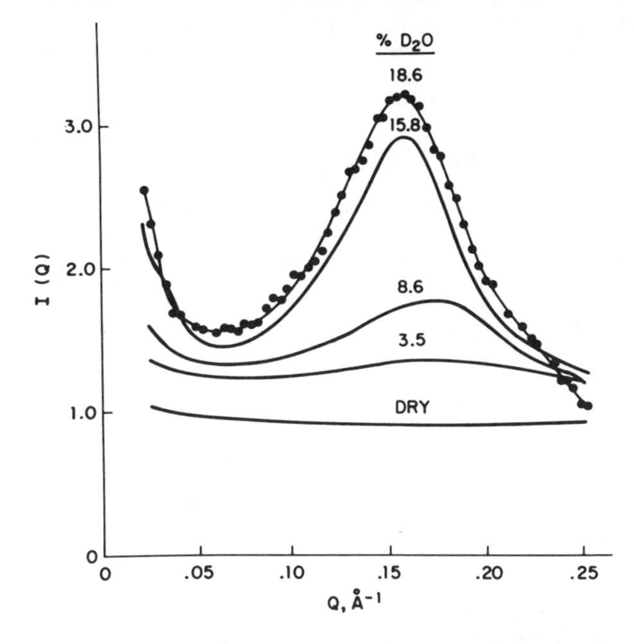

Fig. 3. Neutron scattered intensity vs. scattering vector for a 17 mol% cesium sulfonated polypetenamer. Numbers above each curve indicate weight percent D_2O. (reprinted with permission from ref 20. Copyright 1982 American Chemical Society)

appeared to contain 300nm phase separated regions. Osmium tetroxide staining of these EPDM sections revealed domains averaging less than 3 nm in size primarily inside those regions. Unfortunately, the section thickness prohibited an accurate determination of the size distribution or the detailed shape of these domains.

APPLICATIONS

Ionomers of practical interest have been prepared by two synthetic routes: (a) copolymerization of a low level of functionalized monomer with an olefinically unsaturated monomer or (b) direct functionalization of a preformed polymer. Typically, carboxyl containing ionomers are obtained by direct copolymerization of acrylic or methacrylic acid with ethylene, styrene and similar comonomers by free radical copolymerization. Rees (22) has described the preparation of a number of such copolymers. The resulting copolymer is generally available as the free acid which can be neutralized to the degree desired with metal hydroxides, acetates and similar salts. Recently, Weiss et al.(23-26) have described the preparation of sulfonated ionomers by copolymerization of sodium styrene sulfonate with butadiene or styrene.

The second route to ionomers involves modification of a preformed polymer. Sulfonation of EPDM, for example, permits the preparation of sulfonated-EPDM with a level of sulfonate groups in proportion to the amount of sulfonating agent(27). These reactions are conducted in homogeneous solutions permitting the direct neutralization of the acid functionality to the desired level. Isolation of the neutralized ionomer is effected by conventional polymer isolation techniques, such as coagulation in a nonsolvent or solvent flashing. These procedures are detailed in several patents and publications(28-31).

Ionic Elastomers and Plastics

Over the past 30 years a number of ionic elastomers and plastics have been developed that possess a wide variety of properties leading to different applications. A list of some representative ionomers is given in Table I.

In the early 1950's, B.F. Goodrich introduced the first commercial elastomer based on ionic interactions, a poly(butadiene-co-acrylonitrile-co-acrylic acid). Typically less than 6% of carboxylic monomer is employed in order to preserve the elastomeric properties inherent in these systems. When neutralized to the zinc salt, these elastomers display enhanced tensile properties and improved adhesion compared to conventional copolymers. This enhancement of properties can be directly attributed to ionic associations between the metal carboxylate groups.

A second family of ionic elastomers based on the sulfonation of chlorinated polyethylene was also introduced in the early 1950's by E. I. du Pont de Nemours & Co., Inc. Curing these materials with various metal oxides gives rise to a combination of ionic and covalent crosslinks, and these elastomers are commercially available under the trade name Hypalon.

Table 1 - Representative Ionomers

Polymer System	Trade name, if Commercial	Manufacturer	Uses
Commercial			
Poly(ethylene-co-methacrylic acid)	Surlyn	DuPont	Flexible thermoplastic
Poly(butadiene-co-acrylic acid)	Hycar	BF Goodrich	High green strength[a] elastomer
Telechelic carboxylate elastomers	Hycar	BF Goodrich	Specialty uses
Chlorosulfonated polyethylene	Hypalon	DuPont	Elastomeric Sheeting
Sulfonated ethylene-propylene terpolymer	Ionic Elastomer[b]	Uniroyal	Thermoplastic elastomer
Perfluorosulfonate Ionomers	Nafion	Dupont	Multiple membrane uses
Perfluorocarboxylate Ionomers	Flexion	Asahi Glass	Chloralkali Membrane
Experimental			
Sulfonated butyl elastomer			High green strength[a] elastomer
Thermal reversible SBR[c]			Modified SBR
Sulfonated polypentenamer			Model ionomer
Telechelic polyisobutylene sulfonate ionomers			Model ionomer
Sulfonated polystyrene			Model ionomer
Poly(styrene-co-acrylic acid)			Model ionomer

[a] Green strength = gum tensile strength (prior to vulcanization)
[b] Development stage
[c] (styrene/butadiene/N-isobutoxy methyacrylamide terpolymer ionomers)

In the mid-1960's Dupont introduced poly(ethylene-co-methacrylic acid) under the trade name Surlyn; these were partially neutralized with sodium or zinc cations. These modified polyethylenes possess remarkable clarity and tensile properties superior to those of conventional polyethylene. The high melt viscosity that results from ionic interactions that persist even at elevated temperatures offers advantages in heat sealing and extrusion operations. Other properties attributable to ionic aggregation include toughness, outstanding abrasion resistance, and oil resistance in packaging applications.

A specialty class of carboxyl containing elastomers are the telechelic ionomers. In these systems the carboxyl functionality terminates both ends of the polymer chain. Such polymers range in molecular weight from 1500 to about 6000. These materials can be prepared via several synthetic routes involving anionic or free radical initiated polymerization(32-34). Recently, telechelic sulfonate ionomers of polyisobutylene have been synthesized(35). These systems offer an unusual opportunity to assess the influence of chain length, chain architecture, cation type, and the influence of polar additives on ionomer properties.

Thermally reversible ionomers based on styrene-butadiene rubber, SBR, were prepared by Gunisin(36). The incorporation of functional monomers containing acid, amide, or amine side groups is achieved by emulsion terpolymerization techniques. Usually the amount of the functional monomer is kept low, about 1-3%. Metal or halogen ions are incorporated either by coagulation or bulk mixing. The tensile properties of these polymers are enhanced by ionic associations, and this technique may be used to enhance green strength in elastomers where this characteristic is deficient. Gunisin also reports that blends of an SBR containing small amount of N-(isobutoxymethyl)acrylamide, IBM, salts with plastics such as polystyrene or styrene-acrylonitrile copolymers result in impact strength substantially higher than with systems without IBMA.

A thermopolastic elastomer based on sulfonated-EPDM, S-EPDM, was developed in the 1970's by Exxon and more recently by Uniroyal. Unlike the synthesis of the carboxylate ionomers described above, S-EPDM is prepared by a post-polymerization sulfonation reaction(28). Compared to the metal neutralized S-EPDM, the sulfonic acid derivative is not highly associated. The free acid materials possess low strengths and are less thermally stable. The metal salts of S-EPDM have properties comparable to crosslinked elastomers, but they do exhibit viscous flow at elevated temperatures. In the absence of a polar cosolvent, such as methanol, hydrocarbon solutions of the metal salts of S-EPDM are solid gels at polymer concentrations above several percent(31). With the addition of 1 to 5% alcohol the polymer solution becomes fluid with solution viscosities of the order of 10 to 100 poise.

Most of the published work on ionomers has been concerned with different approaches to incorporation of ionic groups and the resulting influence of these associations on bulk or solution properties. Studies by Makowski et al.(37), Agarwal et al.(38), Weiss(39) and

Duvdevani(40) have been directed at modification of ionomer properties by employing polar additives to specifically interact or plasticize the ionic interactions. This plasticization process is necessary to achieve the processability of thermoplastic elastomers based on S-EPDM. Crystalline polar plasticizers such as zinc stearate can markedly affect ionic associations in S-EPDM. For example, low levels of metal stearate can enhance the melt flow of S-EPDM at elevated temperatures and yet improve the tensile properties of this ionomer at ambient temperatures. Above its crystalline melting point, ca. 120°C, zinc stearate is effective at solvating the ionic groups, thus lowering the melt viscosity of the ionomer. At ambient temperatures the crystalline additive acts as a reinforcing filler.

Weiss, et al.(41) have studied the effect of the counterion on the physical properties of sulfonated polystyrenes with degrees of substitution up to 21 mol%. They neutralized these materials with mono-, di-, and tri-substituted alkyl amines with alkyl chain lengths up to C_{20}. They demonstrated that the ionic interactions significantly weakened as the size of the cation increased. In fact, for extremely bulky substituents such as in tri-stearyl or tri-lauryl amine salts the glass transition temperature and the melt viscosity were much lower than for the unionized polystyrene. This work represents the first case where ionic substituents were used to internally plasticize a polymer. It also demonstrates the extreme flexibility one has to vary the properties of an ionomer by selection of the counterion.

Membranes

Crosslinked polyelectrolyte resins in the form of beads, powders, membranes and coatings are useful in a variety of separations applications, e.g., ion-exchange and electrodialysis(42,43). Over the past 40 years there has developed an extensive literature dealing with the swelling, ion-exchange and transport properties of these materials. The relationship between these properties and factors such as crosslink density, pK and number density of ionogenic groups, nature of counterion, and external solution concentration have been established through exhaustive experimental studies. These developments have been accompanied by molecular or semi-molecular theoretical models of the thermodynamic states of crosslinked polyelectrolyte gels and by continuum mechanical models of the steady-state transport of electrolytes through such gels.

In the last 1960's ionomer membranes became available. In contrast with the traditional ion-exchange resins, ionomers are rendered insoluble by the presence of crystalline domains and ionic clusters. A good measure of past and continuing interest in ionomer membranes issued from the development of perfluorinated ionomers, the first-announced being Nafion(44). These materials are characterized by remarkable chemical resistance, thermal stability and mechanical strength, and they have a very strong acid strength, even in the carboxylic acid form. The functionalities that have been considered include carboxylate, sulfonate, and sulfonamide, the latter resulting from the reactions of amines with the sulfonyl fluoride precursor.

Although perfluorinated ionomers have been evaluated as membranes in a number of applications, such as water electrolysis, fuel cells, air-driers, Donnan dialysis in waste metal recovery, and acid catalysis, the primary system of practical interest is the production of chlorine and caustic by electrolysis. The major advantage of a membrane process is the ability to produce high concentration caustic soda directly without requiring the energy intensive evaporation step. Use of the perfluorinated membrane results in long separator lifetimes, product purity, and high efficiency with low power consumption. The significance of these membranes within the context of chlor-alkali electrochemical cells is discussed by Dotson later in this book.

The economic advantge of chlor-alkali processes based on ionomer membrane technology over more traditional separation processes has spawned considerable research in an effort to develop relationships between the microstructure of the ionomer and the selective transport of ions acorss these membranes. These studies have been directed at the chemistry and physics of ionomer membranes as well as the engineering aspects of their performance. For example, a direct investigation of the operating parameters on the performance of a chlor-alkali cell as been described by Yeager and Malinsky(45). Using a laboratory-scale cell designed for automated operation, they characterized membrane permselectivity and resistance as a function of solution concentration, temperature, and current density.

On the other hand, the nature of the microstructure and the physics of concentrated electrolytes in the context of these systems have also been considered. Hsu(46) has formulated a theoretical percolation model of ion transport that considers ionic clusters that conduct water but which cannot contribute to long-range transport at low water contents where no connectivity of clusters is expected. As the water content increases, an insulator-to-conductor transition occurs at a cluster volume-fraction percolation threshold.

Infrared studies of ionomers have provided information on the cluster and microstructure and the nature of ion-ion, ion-water, and water-backbone interactions(47,48). Risen(49) has extended thee studies into the far-infrared spectral region that is characterized by ion motion vibrational bands that reflect the force constants of cation-anion sidechain interactions. In a later chapter of this book Risen discusses results for Nafion and several other ionomers, and he relates his findings to the energetics of macromolecular organization and the glass transition. From the standpoint of industrial applications of ionic separations using ionomer membranes, these studies may ultimately prove to be important in the understanding the relaxation behavior of ionomers.

Until recently, perfluorinated ionomrs with high equivalent weights were believed to be insoluble. Covitch(50), however, has identified a number of solvents and dissolution procedures for the sulfonyl fluoride precursor and sulfonate and carboxylate Nafion ionomers with 1100 to 1200 equivalent weight. This development has great potential for the preparation of sulfonate and carboxylate ionomer blends, the application of very thin films to electrodes or other substrates, and the production of porous membranes that may be useful in chlor-alkali diaphram cells and solid polymer electrolyte electrodes.

Polyelectrolyte Complexes

Mixtures of oppositely charged polyelectrolytes dissolved in water can interact to form a variety of precipitates, gels, or phase-separated solutions. What is formed depends on the mixing conditions and the density of ionic charges carried by the polymer chains. Polyelectrolytes with high charge densities usually interact to form precipitates. As the charge density decreases, liquid-liquid phase separation, called complex coacervation, occurs.

Complex coacervation was extensively studied between 1930 and 1945 by Bungenberg de Jong and coworkers(51). Although this group focused its studies on the bheavior of gelatin and gum arabic, many mixtures are capable of forming complex coacervates. For example, complex coacervates have been formed from gelatin and pectin, serum albumin and gum arabic, haemoglobin and gum arabic, dupin and gum arabic and histon and DNA. These biopolymers will also form complex coacervates with various synthetic polyelectrolytes.

Complex coacervation is affected by pH, polyelectrolyte concentration, polyelectrolyte mixing ratio, and the neutral salt concentration. It normally occurs over a limited pH range and is suppressed by neutral salts. The optimum polyelectrolyte mixing ratio is that which neutralizes the ionic charges carried by the polyelectrolytes. The coacervation intensity is increased by dilution, which has led some researchers to suggest that complex coacervation is the mechanism by which various polymeric species are selectively isolated and concentrated in biological systems.

Although complex coacervation has been known since the 1930's, the phenomenon was not commercially exploited until 1957. At that time, Green and Schleicher(52) developed a process for fabricating microcapsulates by coacervating gelatin with gum arabic. This process played a key role in the development of carbonless copy paper and led to other studies of the gelatin-gum arabic encapsulation process. As a result, a wide range of hydrophobic materials, such as liquid crystals, flavors, fragrances, vitamins, and organic solvents, can now be microencapsulated with gelatin-gum arabic coacervates.

Because of the success of gelatin-gum arabic microcapsules, encapsulation processes based on other complex coacervation reactions are receiving considerable attention. The driving force for this activity is to reduce the materials costs and to alter the capsule properties by using different polyelectrolytes. Two reactions that have been used successfully are the coacervation of gelatin with polyphosphate(53) and the coavervation of pI 5 gelatin with pI 9 gelatin(54). The gelatin-polyphosphate interactions are more intense than the gelatin-gum arabic interactions, while the gelatin pI5-gelatin pI9 interactions are less intense.

The formation of microcapsules is only one example of the use of polyelectrolyte complexes. Another example is the formation of a permselective membrane around live cells by a complex formed between alginate and poly(L-lysine)(55). This complex is formed under mild conditions so as not to harm the cells, and the membrane can be tailored so that it is permeable to cell nutrients, but impermeable to the cellular products. This process is of interest to those concerned with large-scale cell cultures.

Biopolymers

With the exception of natural polyisoprene, all the important biological macromolecules are polyelectrolytes. Proteins contain weakly acidic and weakly basic groups and their net charge is affected by the pH of the surrounding medium. The nucleic acid backbone incorporates the relatively more acidic phosphate diester linkage in strict alternation with D-ribose or 2-deoxy-D-ribose to yield a chain with a large negative linear charge density over most of the biologically significant pH range. Many naturally occurring polysaccharides are nonionic. The majority are, however, polyelectrolytes with charges arising either from weakly acidic uronate carboxyl groups or from strongly acidic sugar sulfate derivatives; the linear charge densities at neutral pH vary widely depending on the chemical structure. Ionic derivatives of certain neutral polysaccharides, most notably, carboxymethyl cellulose, are of considerable commercial importance.

The ionic character of biopolymers is important in guaranteeing that they are well solvated in the predominantly aqueous media in which they function. Most biological macromolecules are not dissolved (i.e., molecularly dispersed) *in vivo*. Their cellular structures, however, are usually in intimate contact with the ambient aqueous solution of the cytoplasm or intercellular fluid and must be able to readily interact with it.

These interactions are frequently ionic in character. The coulombic forces of interaction between macroions and lower molecular weight ionic species are central to the life processes of the cell. For example, intermolecular interactions of nucleic acids with proteins and small ions, of proteins with anionic lipids and surfactants and with the ionic substrates of enzyme catalyzed reactions, and of ionic polysaccharides with a variety of inorganic cations are all improtant natural processes. Intramolecular coulombic interactions are also important for determining the shape and stability of biopolymer structures, the biological function of which frequently depends intimately on the conformational features of the molecule.

Theoretical considerations of the coulombic interactions of dissolved biopolymers have produced a complete picture of the distributions of counter and coions under the influence of the electrostatic charge on the macroion(56,57). The counterion condensation theory of Manning(56) has stimulated a great deal of activity in the study of dissolved macroions, especially because it provides a group of limiting laws describing the contribution of electrostatic effects to the thermodynamic and transport properties of polyelectrolyte solutions. Data gathered over the past decade confirm the general validity of the limiting laws of counterion condensation theory. An alternative approach, complementary to the counterion condensation theory, has been to solve the classical Poisson-Boltzmann equation(57). When the macroion is modeled as an infinite cylinder, an idealization also employed in counterion condensation theory, limiting laws in agreement with those from the Manning theory are obtained(57). In addition, more direct information about the distribution of small ions in the vicinity of the macroion becomes available.

Several chapters of this book discuss applications and extensions of the theory of polyelectrolyte solutions. Counterion condensation theory postulates that for a cylindrical macroion, if the linear charge density exceeds a well-defined critical value, a sufficient fraction of the counterions will "condense" into the immediate domain of the macroion so as to reduce the net charge density due to the macroion and its condensed counterions to the critical value. No condensation is predicted for macroions with less than the critical charge density.

Counterion condensation theory, however, does not provide a detailed picture of the distribution of the condensed ions. Recent research using the Poisson-Boltzmann approach has shown that for cylindrical macroions exceeding the critical linear charge density the fraction of the counterions described by Manning theory to be condensed remain within a finite radius of the macroion even at infinite polyion dilution, whereas the remaining counterions will be infinitely dispersed in the same limit. This approach also shows that the concentration of counterions near the surface of the macroion is remarkably high, one molar or more, even at infinite dilution of the macromolecule. In this concentrated ionic milieu specific chemical effects related to the chemical identities of the counterions and the charged sites of the macroion may occur.

An important application of polyelectrolyte theory has been to elucidate the role of the tightly held counterions in those conformational changes of biopolymers that may alter the liner charge density of the macroion and thus the numbers of bound counterions(58,59). Because substantial changes in the numbers of bound counterions may accompany a given conformational change, the equilibrium distribution of the system among the several conformational forms of a macroion may be quite sensitive to the added salt concentration. Significant small ion redistribution may also occur upon interaction of two biopolymer molecules. The influence of low molecular weight electrolyte concentration on the equilibria and kinetics of the interactions between nucleic acids and proteins suggests that salts may play an important regulatory role in the protein-nucleic acid interactions involved in gene expression(60). Similarly, the stabilities of ordered structures in ionic polypeptide chains are markedly affected by interactions with small molecules of opposite electrical charge(61). This has implications for the conformational changes induced in certain peptide hormones as a consequence of interactions with anionic lipids of the sort that might occur at or near receptor sites on the cell surface.

The electrostatic potential generated by a dissolved biological macroion is clearly important for determining the equilibrium macromolecular conformation and the distribution of counterions or other ionic ligands in its immediate environment. Recently it has been recognized that the electrostatic field in the vicinity of a biological macroion may influence the dynamics of its biologically important interactions. In particular, analysis of the electrostatic field vectors in the vicinity of the active site of the enzyme Cu, Zn-superoxide dismutase (SOD) suggests that the electrostatic field is instrumental in guiding the negatively charged substrate, the superoxide ion, into the catalytic center(62). Computer graphics techniques have been employed to resolve the contributions to the electrostatic potential gradient of SOD from individual charged residues(63). Similar methods have been used to illustrate the electrostatic potential surfaces of a variety of biological macromolecules and to emphasize electrostatic and steric complementarity in biologically important macromolecule-ligand interactions.

Literature Cited

1. Ionic Polymers, L. Holliday, Ed., Applied Science Publ., London, 1975.

2. A. Eisenberg and M. King, Ion-Containing Polymers, Academic Press, N.Y., 1977.

3. Ions in Polymers, A. Eisenberg, Ed., Advances in Chem. Ser., 187, American Chemical Society, Wash. D.C., 1980.

4. Perfluorinated Ionomer Membranes, A. Eisenberg and H. J. Yaeger, Eds., ACS Symposium Series 180, American Chemical Society, Wash D.C., 1982.

5. Developments in Ionic Polymers-I, A. D. Wilson and H. J. Proser, Eds., Applied Science Publ. London, 1983.

6. E. P. Otocka, J. Macromol. Sci.--Revs. Macromol. Chem., C5(2), 275 (1971).

7. W. J. MacKnight and T. R. Earnest, Jr., J. Polym. Sci., Macromol. Rev., 16, 41 (1981).

8. C. G. Bazuin and A. Eisenberg, Ind. Eng. Chem. Prod. R&D, 20, 271 (1981).

9. W. J. MacKnight and R. D. Lundberg, Rub. Chem. Tech., 57(3), 652 (1984).

10. A. Eisenberg, Macromolecules, 3, 147 (1970).

11. W. Forsman, Macromolecules, 15, 1032 (1982).

12. T. R. Earnest, J. S. Higgins, D. L. Handlin and W. J. MacKnight, Macromolecules, 14, 192 (1981).

13. W. J. MacKnight, W. P. Taggart and R. S. Stein, J. Polym. Sci., Polym. Symp. No. 45, 113 (1974).

14. E. J. Roche, R. S. Stein and W. J. MacKnight, J. Polym. Sci., Polym. Phys. Ed., 18, 1035 (1980).

15. M. Fujimura, T. Hashimoto and H. Kawai, Macromolecule, 15, 136 (1982).

16. D. J. Yarusso and S. L. Cooper, Macromolecule, 16, 1871 (1983).

17. G. Fournet, Acta Cryst., 4, 293 (1951).

18. C. T. Meyer and M. Pineri, J. Polym. Sci., Polym. Phys. Ed., 16, 569 (1978).

19. M. Pineri, R. Duplessix, S. Gauthier and A. Eisenberg, in Ions in Polymers, Advances in Chem. Ser., 187, American Chemical Soc., Wash. D.C., 1980, p. 283.

20. T. R. Earnest, Jr., J. S. Higgins and W. J. MacKnight, Macromolecules, 15, 1390 (1982).

21. D. L. Handlin, W. J. MacKnight and E. L. Thomas, Macromolecules, 14, 795 (1980).

22. R. W. Rees, U.S. Patent 3,322,734, to E. I. Dupont de Nemours & Co., 1966.

23. R. A. Weiss, R. D. Lundberg, and A. Werner, J. Polym. Sci., Polym. Chem. Ed., 18, 3427 (1980).

24. R. A. Weiss, S. R. Turner, and R. D. Lundberg, J. Polym. Sci., Polym. Chem. Ed., 23, 525 (1985).

25. S. R. Turner, R. A. Weiss, and R. D. Lundberg, J. Polym. Sci., Polym. Chem. Ed., 23, 535 (1985).

26. R. A. Weiss, S. R. Turner, and R. D. Lundberg, J. Polym. Sci., Polym. Chem. Ed., 23, 549 (1985).

27. R. D. Lundberg, H. S. Makowski and L. Westerman, in Ions in Polymers, A. Eisenberg, Ed., Adv. Chem. Ser. No. 187, American Chemical Soc., Wash. D.C., 1980, p. 67.

28. H. S. Makowski, R. D. Lundberg and G. S. Singhal, U.S. Patent 3,870,841, to Exxon Res. & Eng. Co., 1975.

29. N. H. Canter, U.S. Patent 3,642,728, to Esso Res. & Eng. Co., 1972.

30. H. S. Makowski, J. Bock and R. D. Lundberg, U.S. Patent 4,221,712, to Exxon Res. & Eng. Co., 1980.

31. R. D. Lundberg and H. S. Makowski, in Ions in Polymers, A. Eisenberg, Ed., Adv. Chem. Ser. No. 187, American Chemical Soc., Wash. D.C., 1980, p. 21.

32. S. E. Reed, J. Polym. Sci. A-1, 9, 2147 (1971).

33. D. N. Schulz, J. C. Sandra and B. G. Willoughby, ACS Symp. Ser, 166, Amer. Chem. Soc., Wash. D.C., 1981, Chap. 27.

34. G. Brozo, R. Jerome and Ph. Teyssie, J. Polym. Sci., Polym. Let. Ed., 19, 415 (1981).

35. Y. Mohajer, D. Tyagi, G. L. Wilkes, R. Storey, and J. P. Kennedy, Polym. Bull., 8, 47 (1982).

36. B. Gunisin, this volume

37. H. S. Makowski, P. K. Agarwal, R. A. Weiss, and R. D. Lundberg, Polym. Preprints, 20(2), 281 (1978).

38. P. K. Agarwal, H. S. Makowski, and R. D. Lundberg, Macromolecules, 13, 1679 (1980).

39. R. A. Weiss, J. Appl. Polym. Sci., 28, 3321 (1983).

40. I. Duvdevani, this volume.

41. R. A. Weiss, P. K. Agarwal and R. D. Lundberg, J. Appl. Polym. Sci., 29, 2719 (1984).

42. F. Helfferich, Ion Exchange, McGraw Hill, N.Y. 1962.

43. J. A. Marinsky, Ion Exchange, V. I., Marcel Dekker, N. Y. 1966.

44. "Nafion" is a registered trademark of E. I. DuPont de Nemours Co.

45. H. L. Yeager and J. D. Malinsky, "Permselectivity and Conductance of Perfluorinated Ionomer Membranes in Chlor-Alkali Electrolysis Process", presented at the Amer. Chem. Soc. Mtg., Philadelphia, Aug. 1984.

46. W. Y. Hsu, "Composite Nature of Ionomers", presented at the Amer. Chem. Soc. Mtg., Philadelphia, Aug. 1984.

47. K. A. Mauritz, C. J. Hora, and A. J. Hopfinger, in "Ions in Polymers", A. Eisenberg, Ed., Advances in Chem. Ser., 187, American Chemical Society, Wash. D.C., 1980, p. 123.

48. M. Falk, in "Perfluorinated Ionomer Membranes", A. Eisenberg and H. L. Yeager, Eds., ACS Symp. Ser., 180, American Chemical Society, Wash. D.C., 1982, p. 139.

49. W. M. Risen, Jr., "Spectroscopic Studies of Ionic Interactions in Ionomers", presented at the Amer. Chem. Soc. Mtg., Philadelphia, Aug. 1984.

50. M. J. Covitch, "Solution Processing of Perfluorinated Ionomers: Recent Developments", presented at the Amer. Chem. Soc. Mtg., Philadelphia, Aug. 1984.

51. H. G. Bungenberg de Jong, in Colloid Science, v. II, H. R. Kruyt, ed., Elsevier, New York, 1949, chap. 8 and 10.

52. B. K. Green and L. Schleicher, U.S. Pat. 2,800,457 (1957).

53. G. Horger, U.S. Pat. 3,872,024 (1975).

54. A. Veis, J. Cohen, and C. Aranyi, U.S. Pat. 3,317,434 (1967).

55. F. Lim and R. D. Moss, J. Pharm. Sci., 70, 351 (1981).

56. G. S. Manning, Acc. Chem. Res., 12, 443 (1979).

57. C. F. Anderson and M. T. Record, Jr., Ann. Rev. Phys. Chem., 33, 191 (1982).

58. M. T. Record, Jr., C. F. Anderson, and T. M. Lohman, Q. Rev. Biophys., 11, 103 (1978).

59. G. S. Manning, Q. Rev. Biophys., 11, 179 (1978).

60. M. T. Record, Jr., "Regulation of the Equilibria and Kinetics of Interactions of Proteins and Nucleic Acids by Inorganic Acids", presented at the Amer. Chem. Soc. Mtg., Philadelphia, Aug. 1984.

61. W. L. Mattice, "Conformational Changes Accompanying the Interaction of Anionic Detergents with Cationic Polypeptides", presented at the Amer. Chem. Soc. Mtg., Philadelphia, Aug. 1984.

62. E. D. Getzoff, J. A. Tainer, P. K. Weiner, P. A. Kollman, J. S. Richardson, and D. C. Richardson, Nature, 306, 287 (1983).

63. P. K. Weiner, J. A. Tainers, E. D. Getzoffs, and P. A. Kollman, "Electrostatic Forces Between Ligands and Macromolecules", presented at the Amer. Chem. Soc. Mtg., Philadelphia, Aug. 1984.

64. F. C. Wilson, R. Longworth, and D. T. Vaughan, Polym. Preprints, 9, 505 (1968).

65. W. J. MacKnight, W. P. Taggart, and R. S. Stein, J. Polym. Sci., Polym. Symp., 45, 113 (1974).

RECEIVED October 21, 1985

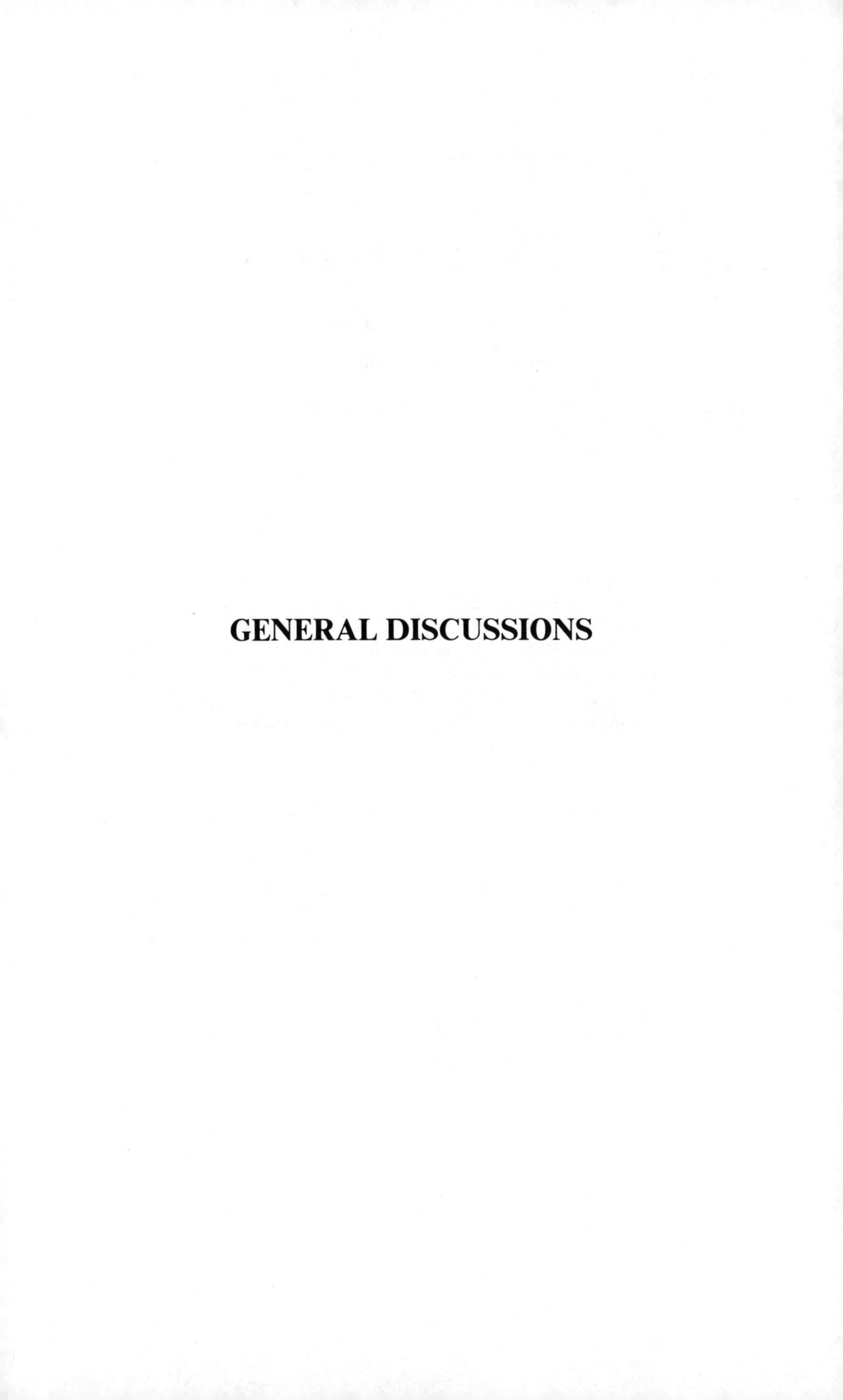

GENERAL DISCUSSIONS

2

Morphological Studies of Model Ionic Polymers

C. E. Williams[1,4], T. P. Russell[2], R. Jérôme[3], and J. Horion[3]

[1]Stanford Synchrotron Radiation Laboratory, Stanford, CA 94305
[2]IBM Research Laboratory, San Jose, CA 95193
[3]Laboratory of Macromolecular Chemistry and Organic Catalysis, University of Liège, Sart Tilman, 4000 Liège, Belgium

Carboxy-halato-telechelic polymers have been studied as models for the more complex ionomers. The effects of well-controlled parameters such as the nature of the metal cation, molecular weight between ionic groups, swelling by polar and nonpolar solvents, ion content and temperature on the size and organization of the aggregates have been investigated by small angle x-ray scattering using a synchrotron radiation source. The principal factor governing the organization of the ionic domains was found to be the root mean square end-to-end distance between the ionic groups. Although aggregation of the ions is dominated by electrostatic interactions, the nature of the cation was found to be only of secondary importance. All the results could be described in terms of ion-multiplets. No evidence was found for cluster formation.

In spite of substantial experimental evidence that aggregation occurs in ion-containing polymers in the bulk, the detailed nature of this aggregation is still unclear. Based on energetic calculations and on an apparent transition in the mechanical and thermodynamic properties with increasing ion content, Eisenberg (1) postulated the existence of two steps in the ion aggregation process: First, *multiplets*, containing not more than eight ion pairs and no organic monomers are formed, driven by an attractive dipolar interaction.

[4]Permanent address: LURE, CNRS, Université Paris-Sud 91405, Orsay, France

0097-6156/86/0302-0022$06.00/0

Over a critical ionic concentration, these multiplets aggregate into larger, ill-defined units called *clusters* which include both ionic and nonionic material. A significant body of experimental data has been rationalized by assuming the existence of these two types of aggregates.

Structural models have also been proposed, based principally on small-angle x-ray scattering (SAXS) results from a large variety of ion-containing polymers. All models consider the multiplet as the basic scattering unit but attribute the origin of the characteristic broad SAXS maximum to either an interference between aggregates (2,3) or to the internal structure of noninteracting particles (4,5). Spherical or lamellar aggregates have been assumed, where the distribution in sizes or separation distances has been taken into account to fit the data in each particular case. However, no clear picture of the aggregates has emerged due to the number of parameters involved in the structural models. In addition, the molecular properties of most of the ionomers studied to date are usually not well defined.

In order to gain some insight into the parameters governing the morphology of ion-containing polymers, we have studied halato-telechelic polymers (HTP) as model compounds for the more complex ionomers. Their molecular properties are well defined and easily varied, as a series of recent articles has shown (6-13). Of critical importance is the fact that the ionic groups are located at the chain ends with a characteric separation distance. These differ from most ionomers where the ion pairs are located on pendant groups randomly spaced along the chain.

We report here a morphological study on a series of HTP using SAXS. In particular, the effect of the metal cation, ion-content, temperature, molecular weight between ionic groups and swelling by polar and nonpolar solvents on the size and organization of the aggregates is examined.

Experimental

Polymer Preparation. Two bifunctional (telechelic) polymers were used in this study. Carboxy-telechelic polybutadiene (PB) is commercially available from B. F. Goodrich (Hycar CTB 2000X156) with molecular characteristics of $\overline{M}_n$=4,600, $\overline{M}_w/\overline{M}_n \simeq 1.8$, functionality$\simeq$2.00 and cis/trans/vinyl ratio of 20/65/15. Carboxy-telechelic polyisoprene (PIP) was prepared by anionic polymerization in THF at −78°C with α-methylstyrene tetramer as a difunctional initiator. The living macrodianions were deactivated by anhydrous carbon dioxide. Five polymers were prepared with $\overline{M}_n$=6,000 10,000, 24,000, 30,000 and 37,000 having $\overline{M}_w/\overline{M}_n \simeq 1.15$; a microstructure ratio of 3, 4/1, 2 of 65/35, respectively, and a functionality >1.95.

The HTPs were then obtained by neutralization of the polymers in solution with highly reactive metal alkoxides or alkyl metals under strictly anhydrous conditions, following a procedure described in detail elsewhere (6,7). When the very rapid reaction is completed, the polymer is dried under vacuum to constant weight. Quantitative neutralization of the carboxylic acid groups can be achieved as evidenced by infrared spectroscopy. Once in the bulk state, all samples were kept under atmospheric conditions and protected from light.

The dried polymers were compression molded into 1-mm thick disks at 100°C and cooled under pressure for 4 hours. When studied in solution, the HTP were examined as prepared. The concentration was adjusted by variation of solvent content. Concentrations were determined by evaporation of the solvent from a known volume of solution and weighing the remaining dried polymer.

Small Angle X-ray Scattering

X-Ray Scattering. All SAXS experiments were performed at the Stanford Synchrotron Radiation Laboratory (SSRL) using the SAXS camera at beam line I-4. Preliminary experiments were conducted at LURE-DCI Orsay at beam line D-11. Both facilities have been described elsewhere (14,15). Briefly, at SSRL, the beam is horizontally focused and monochromated (λ=1.412Å) by a bent, silicon (111) crystal. It is also focused vertically by a float glass mirror so that the beam size at focus several centimeters before the detector is ca. 150×400 μm. Two detectors placed immediately before and after the specimen constantly monitor the beam decay and the sample absorption. The scattering profiles were recorded on a one-dimensional self-scanning photodiode array with a time resolution on the order of 0.5 sec. Two sample to detector distances were used covering a q-range from 0.008 to 0.32$Å^{-1}$ ($q=4\pi/\lambda \sin\theta/2$, where θ is the observation angle and λ is the wavelength). The data corrected for dark current, detector homogeneity, parasitic scattering and sample absorption were matched in the region of overlap. Scattering profiles are presented as normalized intensities *versus* scattering vector q. Conversion to absolute scattering units involves a simple multiplicative factor. Smearing effects due to the dimensions of each pixel in the photodiode array were not taken into account. However, only at very small scattering angles does this become problematic.

Experiments involving temperature variation or swelling were performed in real time as either the temperature was changed or as the swelling solvent was introduced.

Data Analysis

Standard techniques have been used to analyze the SAXS profiles (16). A Guinier analysis of the low q portion of the curves could not be utilized since a plot of $\ell n\, I(q)$ against q^2 was not linear over the q-range investigated. This would have given an estimate of the size of the entities giving rise to the zero-order scattering.

The asymptotic form of the scattered intensity, derived by Porod and Debye for a two-phase system separated by a sharp interface, is given by:

$$\frac{\lim_{q\to\infty} q^4 I(q)}{\int_0^\infty q^2 I(q)dq} = \frac{1}{K_p} = \frac{O}{4\phi(1-\phi)V} \tag{1}$$

where K_p is the Porod constant, O/V is the interfacial area per unit volume and ϕ the volume fraction of one of the phases. The radius of the ionic domains R_D were deduced assuming spherical particles. Electron density fluctuations I_{fl} and diffuse phase boundary thicknesses E_D were derived from the deviation to Porod's law.

The density fluctuations show the presence of ions in the nonionic matrix or of polymer in the ionic microphase. However, the latter contribution is negligible since the volume fraction of the ionic phase is so small.

In some cases, a Debye-Bueche analysis was attempted. In such an analysis, the spatial correlation function is given by:

$$\gamma(r) = \exp(-r/a) \tag{2}$$

where a is a correlation distance. The correlation distance is related to the average size of the heterogeneities or phases but an exact physical interpretation is difficult to assess for concentrated systems.

Experimental Results

Nature of the Cation. Typical SAXS profiles are shown in Figure 1, for two salts of dicarboxylic PB containing 2.3 mol% of acid groups (or percent of charged monomers). All show the broad maximum, centered around $q=0.1Å^{-1}$, characteristic of ionomers and usually taken as evidence for ionic aggregation. This maximum is not seen in the acid form, however. Two other features of importance are the shouldering on the high q side of the main peak, not usually observed in other systems and a scattering at very small scattering angles often neglected in the interpretation of the data. The results summarized in Table I, show that:

- d, the distance between scattering domains derived from Bragg's law applied to the peak, varies little with the nature of the ion and does not seem to be related to its ionic radius.

- 1/d increases with the charge density of the domains $q\rho/m$ (Figure 2) where q, ρ and m are the charge, density and atomic weight of the metal cation, respectively. A asymptote at higher $q\rho/m$ values appears to have been attained. For zero charge density, the data extrapolate to the root mean square end-to-end distance of the PB chain. It should be noted that similar tendencies are observed if the data are plotted as a function cq/a, the parameter introduced by Eisenberg, where c is the anion concentration and a is the distance separating the charges. cq/a relates the critical concentration for clustering and glass transition temperature of the ionomer to the various ions. However, evaluation of the parameter a is not straightforward and involves an assumption on the structure of the multiplets.

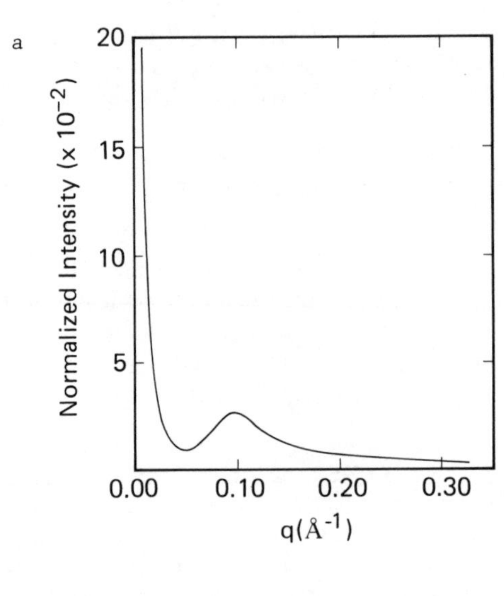

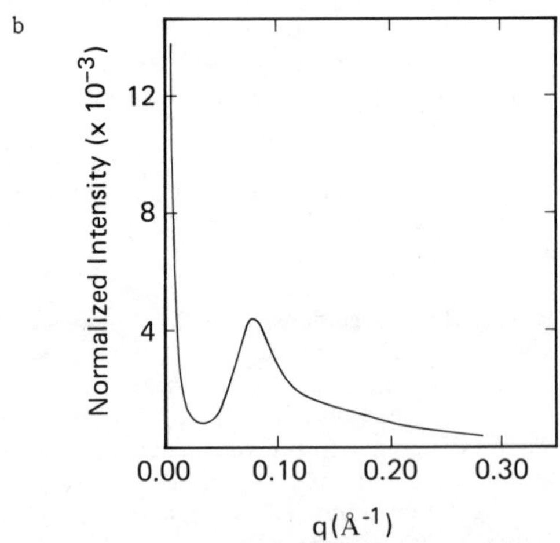

Figure 1. SAXS profiles of dicarboxylic poly(butadiene) having molecular weight 4600 neutralized with (a) Mg^{++} and (b) Ti^{++}.

Table I. SAXS Data for PB with Different Cations

Cation	R_{ionic}	q	d (Å)	R_D (Å)	E_D (Å)
K	1.33	+1	71.6	11.1	5.8
Be	0.34	+2	59.8	9.15	2.6
Mg	0.74	+2	63.0	7.75	0.75
(*)Cu	0.80	+2	54.3	6.2	1.5
Ba	1.38	+2	75.3	6.9	2.0
Fe	0.67	+3	56.1	6.1	1.6
Ti	0.64	+4	57.8	7.65	0.8
Zr	0.82	+4	87.1	8.2	7.0

(*)75% neutralized

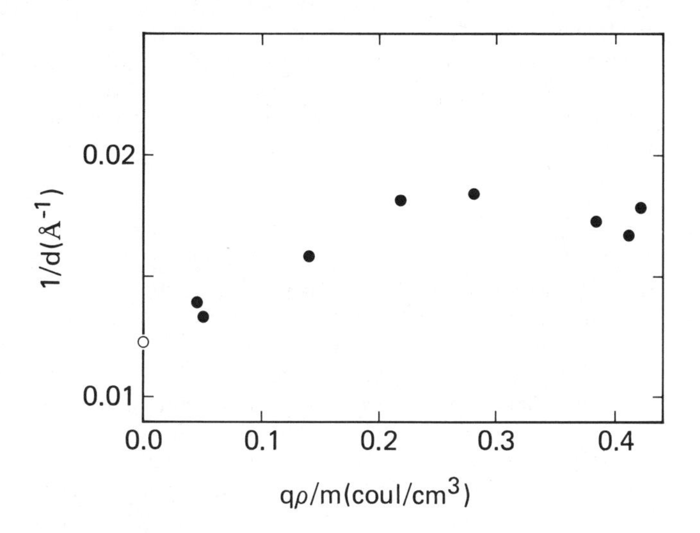

Figure 2. Variation of the reciprocal of the Bragg spacing (1/d) with charge density of the metal cation (qρ/m). The extrapolated value of zero charge density corresponds to the root mean square end-to-end distance of the PB chain.

- In general, the phase boundary, E_D, between the ionic domain and the hydrocarbon matrix appears to be sharp. In fact, the values of E_D are effectively zero to within experimental errors. The only exceptions to this are the K^+ salt, the only monovalent salt studied, and the Zr^{4+} salt.

Molecular Weight Variation. Five samples of carboxylato-polyisoprene 100% neutralized with Mg (PIP-Mg), of $\overline{M}_n$ ranging from 6,000 to 37,000 and low polydispersity, were investigated.

A maximum in the scattering profiles was observed for $\overline{M}_n$=6,000, 10,000 and 24,000, a shouldering on the zero-order scattering for $\overline{M}_n$=30,000 and only a very weak zero-order scattering for $\overline{M}_n$=37,000. It was, therefore, impossible to precisely relate the variation of q_{max} to $\overline{M}_n$, and we can only state that d is *strongly dependent* on molecular weight, *i.e.*, distance between the charges along the chain. It should be noted that the effective ion content also varies with $\overline{M}_n$, where the number of ionizable groups per 100 monomers ranges from 0.37 to 2.12. Since it was impossible to ascertain whether or not aggregation was present in the $\overline{M}_n$=37,000 sample, we did not quantitatively analyze these data.

For the other samples (Table II), the electron density fluctuation scattering for the PIP-Mg series is much lower than for PB-Mg which has a larger polydispersity, indicating that very few ions are outside the ionic domains. The aggregate radius R_D and the diffuse boundary thickness E_D are independent of $\overline{M}_n$ ($R_D \simeq 6$Å; $E_D \simeq 2$Å). The volume fraction of the domains ϕ decreases as $\overline{M}_n$ increases. A Debye-Bueche analysis of the scattering profile shows that the correlation length increases continuously with $\overline{M}_n$. There is no discontinuity in any of the physical quantities measured as a function of $\overline{M}_n$, which would reflect a change in size, shape or organization of the aggregates. Quite simply, as $\overline{M}_n$ increases, the ionic domains appear to be pushed further apart.

Toluene Swelling. All samples swell substantially when immersed in toluene. It is assumed that toluene, a good solvent for polybutadiene, with a low dielectric constant (ϵ=2.38 at 25°C) is absorbed preferentially in the hydrocarbon polymer phase and excluded from the ionic domains. Swelling is then equivalent to changing the distance between domains.

Two different experiments were compared. First, PB-Ti was swollen from the dry state by injecting toluene into the sample cell. The microscopic swelling, measured by the displacement of the SAXS peak (d/d_o) was compared to the macroscopic swelling, measured by a change in linear dimensions of the sample $(V/V_o)^{1/3}$. In a second experiment, solutions were prepared at fixed concentrations, between 7×10^{-2} and 9×10^{-1} g/cm^3. As shown in Figure 3, d/d_o is proportional to $(V/V_o)^{1/3}$ except at very high degrees of swelling. Note that the peak is still visible even when the volume has increased by a factor of 14.

Table II. SAXS Data for PIP-Mg Molecular Weight Series

M_n	R_D^*(Å)	$E_D^†$(Å)	$\phi_{Mg} \times 10^3$	I_{FL}	a(Å)
6,000	5.5	2.0	1.95	30.7	65.6
10,000	6.9	2.0	1.27	11.8	74.8
24,000	5.3	2.7	0.53	16.8	74.2
30,000	5.4	1.1	0.42	28.6	82.3

*R_D values are accurate to within ±15%.

†Values of E_D reported are effectively zero to within experimental errors.

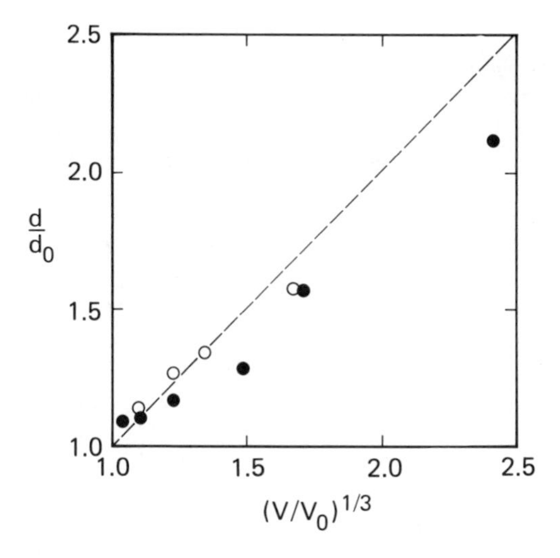

Figure 3. The microscopic swelling (d/d_0) as a function of the macroscopic swelling ($\{V/V_0\}^{1/3}$) for PBTi using toluene as the swelling solvent: ○ samples swollen from the dry state, ● samples prepared in solution. The dashed line corresponds to an affine swelling.

The relative electron density fluctuations I_{FL}/ϕ decreases as the concentration of polymer in toluene ϕ increases. The SAXS invariant, Q, proportional to the volume fraction of the domain, increases with increasing ϕ. R_D and E_D were found to be independent of ϕ. Taken together, these results indicate that when formed, the ionic domains are the same size. However, as the distance between the ionic domains increases, more of the chain ends are forced into the hydrocarbon matrix. This results in a less homogeneous network that can distort as more solvent is added. This is also confirmed by the fact that the breadth of the SAXS reflection relative to the position of the maximum, *i.e.*, $\Delta q/q_{max}$, increases as ϕ decreases. At lower concentrations, the SAXS profiles are skewed towards the higher scattering vectors. This could easily result from an increased contribution to the SAXS from the zero-order scattering. Similar swelling behavior was found for all of the divalent cations studied. It is interesting to note that the materials were somewhat brittle in the swollen state in that they easily fractured during handling. In addition, if the amount of solvent was increased, all of the HTP's would dissolve. These results strongly suggest that the ionic domains (multiplets) acting as the effective crosslinks are not static in structure and have the capacity to rearrange with the application of a strain.

The effects of swelling were reversible provided the swelling ratio (V/V_o) was less than 1.5. Removal of the solvent restored the original structural features of the specimen.

Water Swelling. If water (or any polar solvent) can penetrate the bulk of the polymer, it should be preferentially located in the ionic domains, thereby, creating domains with a high dielectric constant. It could, therefore, produce a morphology of the ionic aggregates different from that of the anhydrous material.

By immersing the sample in water at room temperature for several hours, we found that the PIP-Mg and PB-Ti samples did not swell significantly. However, a very thin layer on the surface became opaque. The SAXS maximum was displaced to lower q values by 7% and 3%, respectively. The most dramatic effects, however, were found for the samples of PB-Mg (Figure 4) neutralized to 80% and 100%. In these specimens, the swelling appeared uniform and the peak sharpened dramatically and was shifted to lower q's by as much as 25%. As expected, the water uptake was accompanied by an increase in domain size from 6.2Å to 11.5Å for PB-Mg (100%) and from 5.8Å to 13.6Å for PB-Mg (80%) with a very slight increase in E_D. The Porod invariant Q and zero-order scattering also decreased.

These results are consistent with an increase in the ionic domain size with water since this would cause a reduction in the domain electron density and, consequently, the total scattering. The invariance of the interphase is surprising, since the domain would be expected to become more diffuse in nature. It is apparent, however, that the hydrocarbon matrix must contain some dissolved ions which are necessary to make the matrix polar enough to support the transport of the water.

Influence of Degree of Neutralization. As was mentioned earlier, a change in the molecular weight between the ionic groups necessitates a change in the ion content of the ionomer. The major consequence of this is to change the average distance between charges along the chain. The ion content may also be varied by changing the degree of neutralization of the diacid PB. This also means that there are carboxyl chain ends in the sample. Figure 5 shows a most striking effect where the d-spacing increases as the degree of neutralization increases from 20% to 100%, following a sigmoidal shaped curve. R_D is independent of the degree of neutralization, I_{FL} stays constant but there is a large decrease in zero-order scattering.

Note that a maximum was observed for all ion contents studied ranging from 0.47 to 2.33 metallic groups per 100 monomers. It is, however, difficult to quantitatively interpret these data since several parameters are simultaneously varying. At 100% neutralization or N=1, all chain ends (except for a few isolated ones) are trapped into the aggregates. As N decreases, more chains are terminated by acid groups which may not necessarily aggregate but may dissolve in the organic matrix. This could change the dielectric constant of the medium in which the aggregates are embedded and, therefore, the ionic interactions. As more chain ends are freed, the network formed by the aggregates becomes looser and the aggregates can come nearer one another. Inhomogeneities in the sample are also more likely to occur in the partially neutralized specimens.

Temperature. Heating the ionomers to 100°C, the annealing temperature, does not significantly change the scattering profile. Above 100°C and up to 200°C, when degradation of the polymer sets in, the zero-order scattering increases, the fluctuation scattering increases and the distance, determined from the peak position, decreases. The maximum is less intense and much broader but still exists at the highest temperature. The effect is much more dramatic for PB-Mg than for PBTi.

It is apparent that a slow melting of the ionic domains occurs above 100°C and more chain ends are forced onto the matrix. The domains become smaller and the distribution of separation distances broadens.

Conclusions

The use of halato-telechelic polymers as model compounds for ionomers has proven useful in gaining some insight into the mechanism of ionic aggregation in these compounds. We can conclude that aggregates exist in all compounds studied in the form of multiplets. This confirms the data on the local environment around the cation obtained by EXAFS on the same samples neutralized with Zn, Ti, Ba, Fe and Cu (13). There is, however, no evidence for the existence of two different types of aggregates or of a critical concentration for clustering. It seems more likely that the cations are either included in the multiplets or isolated in the matrix, in agreement with the conclusions of Yarusso *et al.* on sulfonated polystyrene ionomers (3).

The principal factor governing the organization of the multiplets is the $\overline{M}_n$ of the chain between charges. Furthermore, if the polydispersity is low, as for halato-carboxyl polyisoprene, almost all cations are in the ionic aggregates,

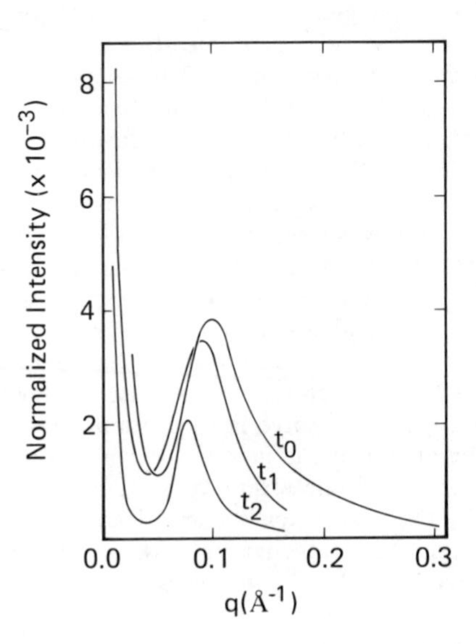

Figure 4. Time evolution of the SAXS for the 100% neutralized PB-Mg swollen with water: (t_0) initial, (t_1) 12 hrs and (t_2) 48 hrs.

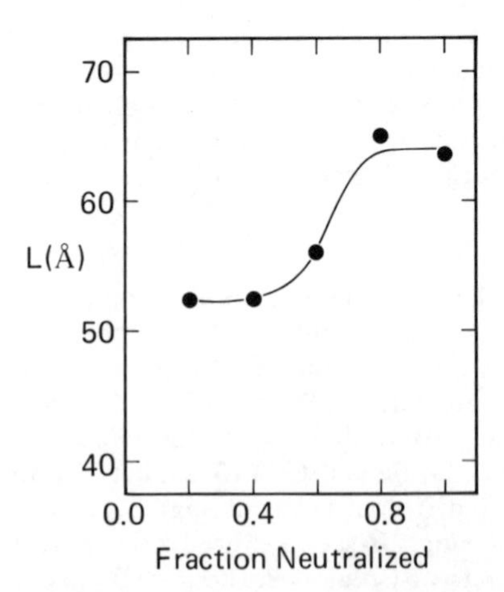

Figure 5. The Bragg spacing as a function of the ion content for PB-Mg.

and there is, as evidenced by the low fluctuation scattering and the lack of water uptake, complete microphase separation between ionic and organic phases.

Although the configuration of the chain between charges could not be determined from the variation of the spacing with molecular weight (both M^1 or $M^{1/2}$ variations were compatible with 3 data points), the amount of swelling observed before disruption of the aggregates is so large that a random coil configuration is likely.

Once formed, the aggregates are very stable (temperature stability, very slow dissolution in polar solvent). If there is complete phase separation, they have a good stability in the presence of humidity, with only a thin surface layer being affected.

Although aggregation is governed by the electrostatic interactions, the nature of the cation is only of secondary importance in the overall organization of the aggregates where only a slight dependence of the separation distance on charge density was observed.

Acknowledgments

We would like to thank Professor Ph. Teyssié of the University of Liège for his interest in this work. One of us (CEW) would like to express appreciation to CNRS, SSRL and IBM France for their financial assistance during the course of this work. We are also grateful to Drs. B. H. Schechtman (IBM San Jose) and A. Bienenstock (SSRL) for their support of this project.

*Some of the materials incorporated in this work were developed at SSRL with the financial support of the National Science Foundation (Contract DMR77-27489) in cooperation with the Department of Energy.

Literature Cited

1. Eisenberg, A., Macromolecules 1970, 3, 147.
2. Marx, C. L.; Caulfield, D. F.; Cooper, S. L. Macromolecules 1973, 6, 344.
3. Yarusso, D. J.; Cooper, S. L. Macromolecules 1983, 16, 1871.
4. MacKnight, W. J.; Taggart, W. P.; Stein, R. S. J. Polym. Sci., Polym. Symp. 1974, No. 45, 113.
5. Roche, E. J.; Stein, R. S.; Russell, T. P; MacKnight, W. J. J. Polym. Sci., Polym. Phys. Ed. 1980, 18, 1497.
6. Broze, G.; Jérôme, R.; Teyssié, Ph. Marcromolecules 1982, 15, 927.
7. Neutralization by Ti or Zr salts, which have a pronounced covalent character, is done in air, the presence of humidity being necessary to form a crosslinked material. See Broze, G.; Jérôme, R.; Teyssié, Ph. J. Polym. Sci., Polym. Letts. 1983, 21, 237.
8. Broze, G.; Jérôme, R.; Teyssié, Ph. Macromolecules 1981, 14, 224.
9. Broze, G.; Jérôme, R.; Teyssié, Ph.; Marco, C. Polymer Bull. 1981, 4, 241.
10. Broze, G.; Jérôme, R.; Teyssié, Ph.; Gallot, B. J. Polym. Sci., Polym. Lett. Ed. 1981, 19, 415.
11. Broze, G., Jérôme, R.; Teyssié, Ph. Macromolecules 1982, 15, 920; 1300.

12. Broze, G.; Jérôme, R.; Teyssié, Ph.; Marco, C. Macromolecules 1983, 16, 996; 1771; J. Polym. Sci., Phys. Ed., 1983, 21, 2265.
13. Jérôme, R.; Vlaic, G.; Williams, C. E. J. Physique Letts. 1983, 44, L717.
14. Stephenson, G. B., Ph.D. thesis, Stanford University (1982).
15. Tchaubar, D.; Rousseaux, F.; Pons, C. H.; Lemonnier, M. Nucl. Instr. Meth. 1978, 152, 301.
16. Glatter, O; Kratky, O. In "Small-Angle X-Ray Scattering"; Academic Press, 1982.

RECEIVED June 10, 1985

3

Cation–Anion and Cation–Cation Interactions in Sulfonated Polystyrene Ionomers

Spectroscopic Studies of the Effects of Solvents

J. J. Fitzgerald and R. A. Weiss

Institute of Materials Science, University of Connecticut, Storrs, CT 06268

The vibrational modes of the sulfonate group in lightly sulfonated polystyrene ionomers are affected by changes in its local environment. In dry materials, the counterion imposes a strong electrostatic field on the sulfonate anion that polarizes the S-O dipole and shifts the asymmetric and symmetric vibrations to higher frequencies. The addition of solvent weakens the polarization by the cation and shifts the vibrational frequencies to lower frequencies. These results are explained in terms of a solution shell around the cation.

ESR analysis of Mn-SPS and Cu-SPS showed that in the bulk the cations are primarily associated. In solution the ratio between associated and isolated cations depends on the polarity of the solvent. The more polar the solvent, the greater is the extent of isolation of the cation.

Ionomers are of considerable technological and scientific interest due to the unique mechanical and rheological properties that arise from intermolecular interactions of the ionic moieties. The literature concerning ionomers is now quite extensive and several recent reviews(1-3) and books(4,5) discuss the synthesis, structure, and properties of these materials.

Surprisingly few studies have focused on the effect of solvents or diluents on the structure and properties of ionomers. Solution results are scarce due to the limited solubility of ionomers in conventional solvents, because of the strong intermolecular associations of the ionic groups(6,7).

Very few studies have considered the behavior of ionomers in relatively polar solvents, i.e., solvents with high dielectric constants, ε. Schade and Gartner(8) compared the solution behavior of ionomers derived from copolymers of styrene with acrylic acid, methacrylic acid, or half esters of maleic anhydride in tetrahydrofuran (THF), a relatively non-polar solvent ($\varepsilon = 7.6$), and dimethyl formamide (DMF), a polar solvent ($\varepsilon = 36.7$). They ob-

0097–6156/86/0302–0035$06.00/0

served a polyelectrolyte effect in dilute solutions of the sodium (Na^+) salts in DMF, but not for the magnesium (Mg^{2+}) salts. The authors concluded that in the case of the Na-salt in DMF the carboxyl group is fully ionized and the increase in the reduced viscosity at low polymer concentrations (the polyelectrolyte effect) is a result of elongation of the polymer chain due to intrachain electrostatic repulsions between the ionized anions. For divalent cations such as Mg^{2+} ion-pairs are favored, and intramolecular associations of the carboxylate groups occur in very dilute solutions. When THF was used as the solvent no dissociation of the cation and the anion occurred and ion-pair interactions predominated for both salts.

Similar solution behavior was reported(9-11) for sulfonate ionomers. Rochas et al.(9) observed a polyelectrolyte effect for acrylonitrile-methallylsulfonate copolymers in DMF. Lundberg and Phillips(10) studied the effect of solvents, with dielectric constants ranging from $\varepsilon = 2.2$ to $\varepsilon = 46.7$, on the dilute solution viscosity of the sulfonic acid and Na-salt derivatives of sulfonated polystyrene (SPS). For highly polar solvents such as DMF and dimethylsulfoxide (DMSO, $\varepsilon = 46.7$) they observed a polyelectrolyte effect, but for relatively non-polar solvents such as THF and dioxane ($\varepsilon = 2.2$) no polyelectrolyte effect was observed. Like Schade and Gartner, these authors concluded that polar solvents favor ionization of the metal sulfonate group while non-polar solvents favor ion-pair interactions.

Weiss et al.(11) observed similar effects of solvent polarity on the solution behavior of copolymers of styrene and sodium styrene sulfonate and the Na-salt of an SPS with a lower backbone molecular weight than that studied by Lundberg and Phillips. Weiss and co-workers observed no such behavior with a mixed solvent of toluene ($\varepsilon = 2.4$) and methanol ($\varepsilon = 32.6$). In light of the differences in the solution behavior of Na^+ and Mg^{2+} salts of the carboxylate ionomers in DMF reported by Schade and Gartner, it is worth noting that neither Weiss et al. nor Lundberg and Phillips reported data in polar solvents for cations other than Na^+.

Several spectroscopic studies(12-14) considered hydration of perfluorosulfonate ionomers ("Nafion"). Lowry and Mauritz(12) and Falk(13) used infrared spectroscopy to study the effects of water and counterion on the sulfonate anion. They observed that water shifts the $-SO_3^-$ symmetric stretching vibration to lower frequency and narrows the absorbance band. The magnitude of the shift decreases with increasing cation radius, and the shift begins at lower water to ionic group ratios for the heavier cations. The authors explained this in terms of a multistage association-dissociation equilibrium between bound and unbound cations. Ion pair formation tends to polarize the S-O dipole due to the strong electrostatic field of the adjacent counterions. Hydration of the ionomer shifts the equilibrium to greater dissociation. This is accompanied by a shift in the SO_3^- stretching vibrations to lower frequency which corresponds to less polarization of the S-O dipole.

Komoroski and Mauritz(14) studied the hydration of Nafion ionomers by ^{23}Na-NMR, and they concluded that the interactions of the Na^+ counterions with the sulfonated resin can be treated in terms of an equilibrium between strongly associated ions and relatively free ions. At low water concentrations these authors observed a large chemical shift of the ^{23}Na resonance that they attributed to the formation of contact ion-pairs. At higher water concentrations they observed that the majority of ions were dissociated hydrated ion pairs.

The work described in the present paper concerns the influence of water and organic solvents on the ionic interactions in lightly sulfonated polystyrene (SPS) ionomers. The focus will be specifically directed towards the influence of the solvent environment on the cation-anion and cation-cation interactions. Fourier transform infrared spectroscopy (FTIR) was used to probe the former while electron spin resonance spectroscopy (ESR) was used to study the latter. Experiments were carried out with dissolved, swollen, and bulk ionomers.

Experimental Section

Sample Preparation

The starting polystyrene homopolymer was obtained from Polysciences, Inc., and had number-average and weight-average molecular weights of 20,000 and 203,000, respectively, as determined by gel permeation chromatography. SPS was prepared using the method outlined by Makoswki and Lundberg(15). The sulfonated polymers were neutralized in toluene/methanol solutions (90/10 v/v) with excess sodium hydroxide, manganese acetate, copper acetate or zinc acetate to form the sodium, manganese, copper, and zinc salts, respectively. The resulting ionomers were recovered by steam distillation, washed with methanol and dried for at least 48 hours under vacuum at 90°C.

FTIR Spectroscopy

Thin films, 25 to 76 μm thick, were prepared by compression molding at 180°C. A Nicolet 60SX Fourier transform infrared (FTIR) spectrometer was used to obtain the infrared spectra. One hundred scans were taken with 2 cm^{-1} resolution. Hydrated samples were parepared by suspending a film above distilled water at 25°C in a closed vial.

ESR Spectroscopy

Electron spin resonance (ESR) measurements were made with a Varian E-3 spectrometer operating at an X-band frequency. Samples were prepared by compression molding as previously described. A quartz solution cell made by Wilmad Glass Co. was used for measurements on solutions.

Results and Discussion

FTIR Spectroscopy (Background)

Liang and Krimm(16) discussed the infrared spectrum of polystyrene, summarizing the frequency, relative intensity and origin of each absorbance band. These assignments were based on symmetry considerations of a monosubstituted benzene ring.

After sulfonating polystyrene the symmetry of the monosubstituted benzene ring is altered, and this affects the in-plane stretching vibrations of the benzene ring. In addition, new absorbances arise from the asymmetric and symmetric stretching vibrations of the sulfonate group. Several previous studies have described the infrared absorbances of the sulfonate group in organic salts(17,18), polyelectrolytes(19), Nafion(12,13) and sulfonated polypentamers(20). The book by Zundel(19) gives an especially thorough description of the infrared spectrum of fully sulfonated polystyrene and the effects of the counterion and hydration on the frequency of the sulfonate absorbance bands.

Infrared bands at 1128 cm^{-1} and 1097 cm^{-1} in sulfonated polystyrene result from the in-plane skeleton vibrations of a disubstituted benzene ring and are strongly affected by the second substituent. An absorbance at 1128 cm^{-1} is observed only in hydrated samples. In the hydrated form, the proton of the sulfonic acid is removed from the anion, yielding the sulfonate anion, SO_3^-, which has three identical S—O bonds that resonate between single and double bond character. In the dry form, the sulfonic acid, $-SO_3H$ group, exhibits an infrared absorbance at 1097 cm^{-1}. Another band occurs around 1000 cm^{-1} and is dependent on the degree of hydration. In a dry sample this absorbance band is observed at 1011 cm^{-1} and shifts to lower wavenumber as the sample becomes progressively hydrated. In the hydrated sample an absorbance at 1034 cm^{-1} is present, which is due to the symmetric stretching vibration of the $-SO_3^-$ anion. The SO_3H group exhibits absorbance bands at 1350 cm^{-1}, 1172 cm^{-1} and 907 cm^{-1} when the sample is dry. The absorbance at 1172 cm^{-1} results from the symmetric stretching vibration of two S=O bonds while the band at 907 cm^{-1} is due to the stretching vibration of the S-O(H) bond. The absorbance band at 1350 cm^{-1} is due to the asymmetric stretch of two S=O bonds.(19)

The local structure of the SO_3^- ion is tetrahedral and has C_{3v} symmetry. The asymmetric stretching vibration of this ion is doubly degenerate and when the sample is neutralized the electrostatic field associated with the cation polarizes the anion and changes the symmetry of the SO_3^- anion to C_s symmetry. The degeneracy is removed and the SO_3^- asymmetric stretch splits into a doublet that absorbs at around 1200 cm^{-1} while the symmetric stretching vibration of the SO_3^- ion absorbs around 1040 cm^{-1}. The degree of splitting is dependent on the strength of the electrostatic field generated by the cation and, to a much lesser extent, the cation mass.(19)

FTIR Spectroscopy of SPS (Dry Samples)

The infrared spectra of polystyrene, a 7.6 mole% SPS-acid derivative (H-SPS), and a 1.7 mole % Na-SPS are shown in figure 1. The SPS samples were compression molded at about 200°C and dried for 96 hours at 90°C under vacuum. The IR absorbances resulting from sulfonation and/or neutralization of the free acid and the band assignments are summarized in Table I.

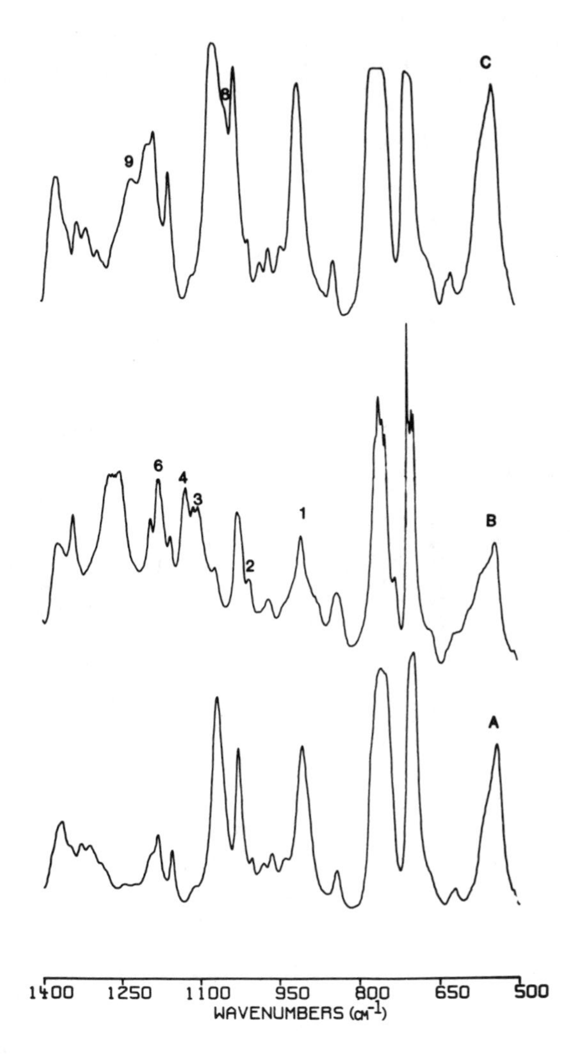

Figure 1. FTIR spectra of compression molded films of A) Polystyrene B) H-SPS (7.6 mole %) C) Na-SPS (1.7 mole %).

Table I - Infrared Absorption Peaks and Assignments[1]

Absorbance Band[2]		H-SPS(7.6%)		Zn-SPS(7.6%)		Mn-SPS(7.6%)		Cu-SPS(7.6%)		Na-SPS(1.7%)
		Dry	Wet	Dry	Wet	Dry	Wet	Dry	Wet	Dry
1	S-O	907								
2	O -S-OH(M^+) O	1011 (s)	1006 (s)	1007	1007 (s)	1009 (s)	1009 (s)	1008 (s)	1008 (s)	
3	O -S-OH O	1101 (s)								
4	SO_3^-	1125 (s)	1125 (s)							
5	SO_3^-		1034 (s)							
6	O=S=O[3]	1176 (s)								
7	O=S=O[4]									
8	$SO_3^-M^{+3}$			1042 (m)	1034 (m)	1048 (s)	1035 (s)	1040 (s)	$\overline{1034}$ (s)	1048 (s)
9	$SO_3^-M^{+4}$			1232	1220	$\overline{1220}$ (s)	$\overline{1212}$ (s)	$\overline{1235}$ (s)	$\overline{1225}$ (s)	1226 (s)
10	$SO_3^-M^{+4}$			$\overline{1272}$ (s)		$\overline{1265}$ (m)		$\overline{1277}$ (s)	$\overline{1275}$ (s)	

[1] Wavenumbers (cm^{-1}), bar denotes broad band: (w) weak, (m) moderate, (s) strong. [2] Numbers correspond to those in figures 1-7. [3] Symmetric stretching vibration. [4] Antisymmetric stretching vibration.

The identification of the sulfonic acid and sulfonate IR bands is complicated by the fact that there are a number of polystyrene absorbances in the same region in which these absorbances occur. Therefore, band assignments were made not only on the basis of new absorbances, but also from changes in the intensity and breadth of previously existing bands. For example, the band at 1176 cm^{-1} in the H-SPS is more intense than the absorbance band at 1182 cm^{-1} in polystyrene. Therefore, the peak at 1176 cm^{-1} may be the result of contributions from the symmetric stretch of two S=O bonds and the original absorbance from the polystyrene. Similarly, the absorbance at 907 cm^{-1} in the H-SPS is much broader than that in polystyrene and may include contributions from an S-O bond. New absorbance bands appear at 1125 cm^{-1} and 1102 cm^{-1}, which indicate that the benzene ring is disubstituted. The absorbance band at ∿838 cm^{-1} indicates that the sample is sulfonated in the para position of the benzene ring.

The 1125 cm^{-1} band, which is due to a sulfonate anion ($-SO_3^-$) attached to the benzene ring, indicates some hydration of this sample. However, not all the sulfonic acid groups are hydrated. This is evident from the 1005 cm^{-1} band (note that this absorbance is more intense and broader than the 1003 cm^{-1} band in polystyrene), which is due to unhydrated sulfonic acid groups, $-SO_3H$. The IR band at 1102 cm^{-} also confirms the presence of $-SO_3H$ groups.

For the 1.7 mole % Na-SPS, the absorbance at 1048 cm^{-1} is probably due to the symmetric stretch of the SO_3^- anion, and the broad band at 1226 cm^{-1} is attributed to the asymmetric stretch of the SO_3^- anion. The second absorbance due to the asymmetric stretch of the SO_3^- anion is not easily identified, but it appears that the 1197 cm^{-1} band in the sulfonated polymer may include a contribution from this vibration (note the difference between the ratio of the 1197 cm^{-1} and the 1155 cm^{-1} intensities in NaSPS and polystyrene).

Divalent transition metal cations such as Cu^{2+}, Zn^{2+}, and Mn^{2+} show different effects presumably because the d-electrons interact very strongly with the SO_3^- anions. The IR spectra of the Cu^{2+}, Mn^{2+} and Zn^{2+} salts of a 7.6 mole % SPS are shown in figure 2. These samples were dried at 90°C under vacuum for 48 hrs.

For the Zn^{2+} salt, the broad absorbance between 1228 and 1240 cm^{-1} is due to the asymmetric stretch of the SO_3^- anion. The symmetric stretch of the SO_3^- anion appears as a shoulder at 1042 cm^{-1}. A band at 1009 cm^{-1} results from the in-plane bending vibrations of a benzene ring substituted with a metal sulfonate group.

The IR spectrum of the Mn^{2+} salt shows a broad asymmetric stretch of the SO_3^- anion between 1210 cm^{-1} and 1230 cm^{-1}. The symmetric stretch of the SO_3^- anion occurs at 1047 cm^{-1}, and a strong absorbance is again noted at 1009 cm^{-1}. The Cu^{2+} salt exhibits a broad absorbance between 1227 and 1239 cm^{-1} that may be attributed to the asymmetric vibration of the SO_3^- anion while the symmetric stretch occurs at around 1040 cm^{-1}. The in-plane skeleton vibration of the benzene ring occurs at 1009 cm^{-1}. An absorbance at 1270 cm^{-1} was observed in the IR spectrum of each transition metal salt, and although the origin of this band is not clear, it may also be due to the asymmetric stretch of the SO_3^- anion. The broadening of the IR spectra in the region of the asymmetric stretching may be a consequence of polarization of the sul-

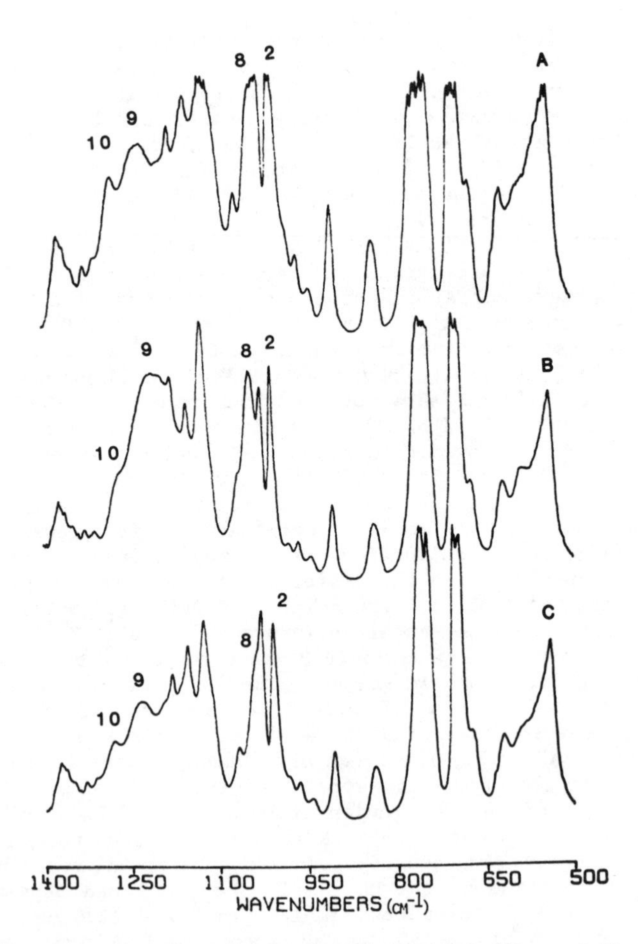

Figure 2. FTIR Spectra of compression molded films of A) Cu-SPS B) Mn-SPS C) Zn-SPS (all 7.6 mole %).

fonate anion by the metal cation. Similar to the explanation of spectral broadening in dry Nafion samples by Lowry and Mauritz(12), the broadened region in SPS may include a number of unresolved peaks arising from a multistage association-dissociation equilibrium between bound and unbound cations.

FTIR Spectroscopy (Hydration and Solvent Studies)

The effect of hydration on the infrared spectrum of a 7.6 mole % H-SPS is shown in figure 3. After 19 hours the sample absorbed about 9% water, which corresponds to 9 moles H_2O per sulfonic acid group. This was accompanied by a large decrease in the intensity of the absorbance band at 1176 cm^{-1} and increases in the intensities of the bands at 1126 cm^{-1} and 1007 cm^{-1}. This result was expected, since upon hydration the acidic proton of the sulfonic acid is removed from the anion (1176 cm^{-1} is characteristic of the S=O symmetric stretch of the $-SO_3H$ and 1126 cm^{-1} and 1007 cm^{-1} are due to the in-plane skeleton and the in-plane bending vibrations of a benzene ring with a SO_3^- group attached). An absorbance at 1033 cm^{-1} in the 19 hour sample is due to the symmetric stretch of the $-SO_3^-$ and is further evidence of hydration. Similarly, the decrease in intensity of the band at 1100 cm^{-1} in the hydrated sample indicates that fewer $-SO_3H$ groups are present.

Further hydration of the H-SPS sample had little influence on the IR spectrum. The spectra given in figure 3 for samples hydrated for 43 and 115 hours are similar to that obtained after 19 hours hydration. This result is consistent with the observation by Lowry and Mauritz(12), that the IR spectrum of Nafion was unaffected by increasing water concentration once a critical hydration occurred.

The effect of hydration on the IR spectrum of the 7.6 mole % Zn-SPS is shown in figure 4. For the sample dried at 90°C under vacuum for 48 hours the band centered at about 1232 cm^{-1} is due to the asymmetric stretch of the SO_3^- anion. An intense absorbance is visible at 1128 cm^{-1} due to the in-plane skeleton vibrations of the benzene ring perturbed by the presence of the SO_3^- anion substituent. Another intense band due to the S—O symmetric stretch of the metal sulfonate group appears at 1042 cm^{-1}.

After the sample was hydrated to about 3.5% (19 hours), which corresponds to 3.6 moles H_2O per sulfonate group, water absorbances were observed in the IR spectrum at 3550 cm^{-1}. The broad absorbance band between 1154 cm^{-1} and 1230 cm^{-1} is difficult to resolve, but the peak near 1270 cm^{-1} in the dry sample disappeared in the hydrated sample and the peak at 1232 cm^{-1} moved to 1220 cm^{-1}. This suggests that the cation is removed from the anion and that the $-SO_3^-$ anion progressed towards its degenerate state (the asymmetric stretch of the degenerate state of the SO_3^- anion occurs at about 1200 cm^{-1}). Further evidence for this is the behavior of the SO_3^- symmetric stretch which shifted from 1042 cm^{-1} to 1036 cm^{-1} upon hydration. (The symmetric stretch of the SO_3^- anion in its degenerate state occurs at 1034 cm^{-1}). No other changes in the IR spectrum occurred upon further hydration.

The IR spectrum of the hydrated 7.6 mole % Mn-SPS, was difficult to interpret due to the broadening of the spectrum between 1160 cm^{-1} and 1235 cm^{-1}. However, the symmetric stretching of the

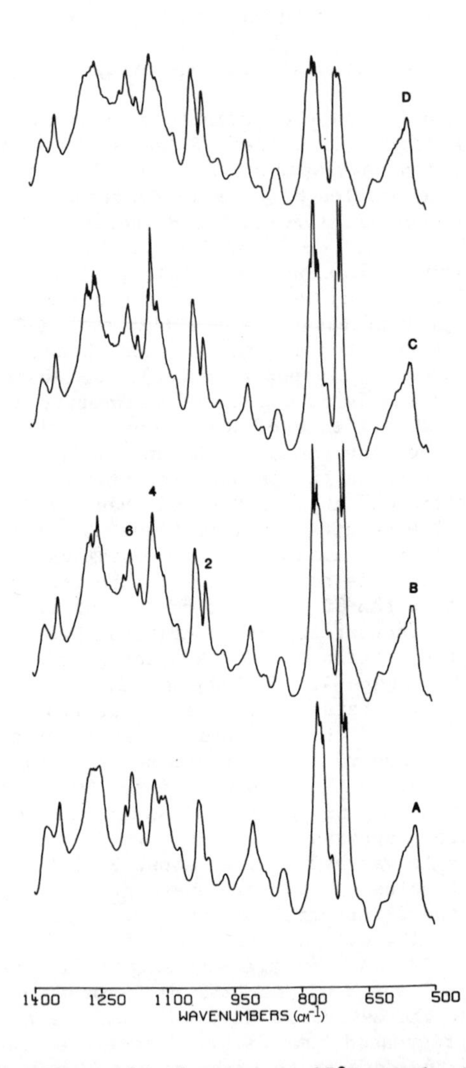

Figure 3. The effect of hydration at 25°C on the FTIR spectrum of H-SPS (7.6 mole %). A) Dry sample, B) Hydrated 19 hours C) 43 hours D) 115 hours.

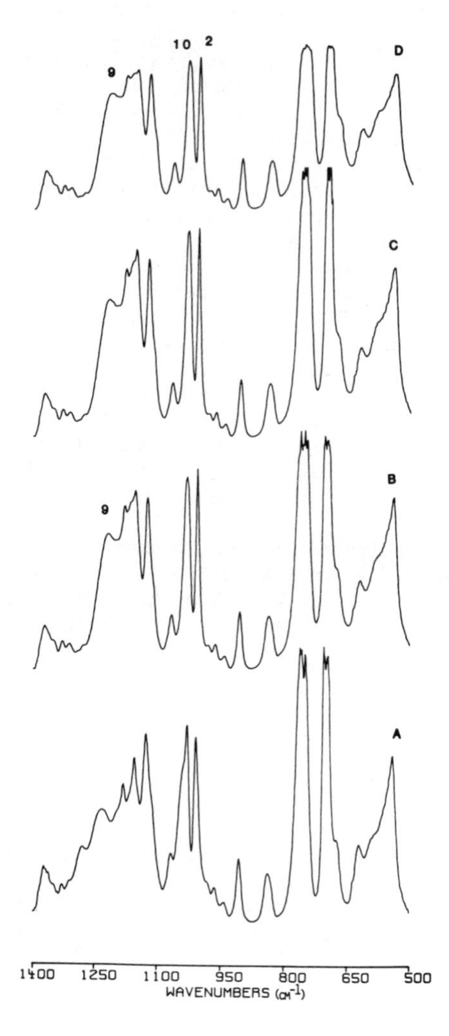

Figure 4. The effect of hydration on the FTIR spectrum of Zn-SPS (7.6 mole %) A) Dry sample B) hydrated 19 hours C) 43 hours D) 115 hours.

SO_3^- moves from 1047 cm^{-1} to 1034 cm^{-1} upon hydration indicating dissociation of the cation and anion. The effect of hydration on the IR spectrum of the 7.6 mole % Cu-SPS, was similar to that of the other two salts. The asymmetric stretching band was difficult to interpret due to broadening but the symmetric stretch of the SO_3^- anion moved from about 1040 cm^{-1} to 1034 cm^{-1}. After 19 hours of hydration no further changes of the spectrum was observed.

In their study of the hydration of Nafion, Lowry and Mauritz(12) concluded that water forms a hydration shell around the cation. This weakens the interaction between the cation and anion to the point where the electrostatic field of the cation has little influence on the stretching vibrations of the anion. At this point the three S—O bonds of the SO_3^- anion are identical and a single asymmetric stretch and a symmetric stretch should occur in the IR spectrum at about 1200 cm^{-1} and 1034 cm^{-1}, respectively(19).

Lowry and Mauritz(12) observed a narrowing of the SO_3^- symmetric stretch in Nafion after hydration, while the spectrum in Figure 4 shows that in SPS the asymmetric stretch broadens. This result is not inconsistent, and both may be rationalized in terms of the ionic-hydrate association model proposed by Lowry and Mauritz. The difference between the results in the two studies may be due to differences in the ability of water to dissociate a transition metal salt as opposed to an alkali metal salt from the sulfonate anion. The broad spectra for the hydrated M^{2+}-SPS samples may be indicative of a number of different ionic-hydrate states.

The effect of solvents on the interactions between the cation and the anion for a 7.6 mole % Na-SPS in THF, DMF, DMSO, and a mixed solvent of 95% toluene and 5% methanol (w/v) is summarized in Table II. The samples used were gels containing about 50% polymer and in each case the solvent spectrum was subtracted from the solution spectrum.

Table II

Effect of Solvent on the Stretching Vibrations of the Sulfonate Anion in 7.6 Mole % Na-SPS

Solvent	Symmetric Stretch (cm^{-1})	Asymmetric Stretch (cm^{-1})
No solvent	1048	1226
THF	1036	1215
toluene/methanol (95/5)	1033	1209
DMF	1034	1217
DMSO	--	1216

The effects of each solvent were similar. The stretching vibrations of the SO_3^- anion decreased, which indicates that the solvent weakened the bond between the cation and anion. It is surprising that in the context of this experiment, relatively nonpolar THF appears to be as effective as polar DMF and DMSO at separating the Na^+ from the SO_3^- anion. Furthermore, the toluene/-methanol mixed solvent does this the most effectively. This last observation is probably a result of a preferential partitioning of the polar methanol to the ionic species.

ESR Spectroscopy

Electron spin resonance (ESR) is useful for probing the local interaction between paramagnetic cations. For example, both Mn^{2+} and Cu^{2+} ions exhibit distinctly different spectra depending upon whether the ions are isolated from each other or associated(21-24).

Isolated Mn^{2+} ions in dilute solution exhibit six hyperfine lines due to coupling of the electron and nuclear spin(21). In more concentrated solutions the Mn^{2+} ions interact more frequently, which results in a broadening of the ESR spectrum. In sufficiently concentrated solutions the hyperfine structure may no longer be resolved and the spectrum becomes a single broad line.

Toriumi et al(24) found that the ESR spectra of solid state Mn-SPS samples with varying sulfonate concentration could be reconstructed from a linear combination of two spectra: one purportedly due to entirely isolated Mn^{2+} ions and one due to entirely associated ions. By doing so, these authors were able to estimate the relative concentrations of the two Mn^{2+} states. For the Cu^{2+} salt of a butadiene-methacrylic acid ionomer, Pineri et al(25) observed two distinct ESR signals: one at 3385G and another at 4735G (f = 9.56GHz). By varying the Cu^{2+} concentration in the ionomer, these authors were able to assign the lower field signal to isolated Cu^{2+} ions and the higher field signal to Cu^{2+} --- Cu^{2+} associations.

ESR spectra of a 2.6 mole % Mn-SPS in bulk and in 5% (w/v) solution in THF and DMF are shown in figure 5. The solid state spectrum exhibits one broad absorption indicating that the population of associated cations is high. The spectrum for the THF solution shows some poorly resolved hyperfine structure. This indicates that both isolated and associated Mn^{2+} ions exist in the THF solution, but the latter species dominate the spectrum. The DMF solution exhibits a classical Mn^{2+} solution spectrum. The well resolved six hyperfine lines indicate that the Mn^{2+} ions are not interacting - that is, there is no perturbation of the electron wavefunction of one cation with the nuclear spin of another cation.

For a toluene/methanol solution, figure 6, the hyperfine structure is well resolved, but the spectrum is broadened as a result of dipole-dipole interactions. As with the THF solution, both isolated and associated ions are present, but in toluene/-methanol the isolated species is predominant. Figure 6 compares the solution spectra of a 1.15 mole % Mn-SPS in toluene with various concentrations of methanol co-solvent. These materials are insoluble in toluene and some methanol is needed in order to dissolve them. At the lowest methanol concentration studied, 2% (v/v), a single broad line spectrum was observed. At higher concentrations of methanol, the six hyperfine structure became evident. These results indicate that the methanol is effective at solvating the cations, which is consistent with the conclusion made

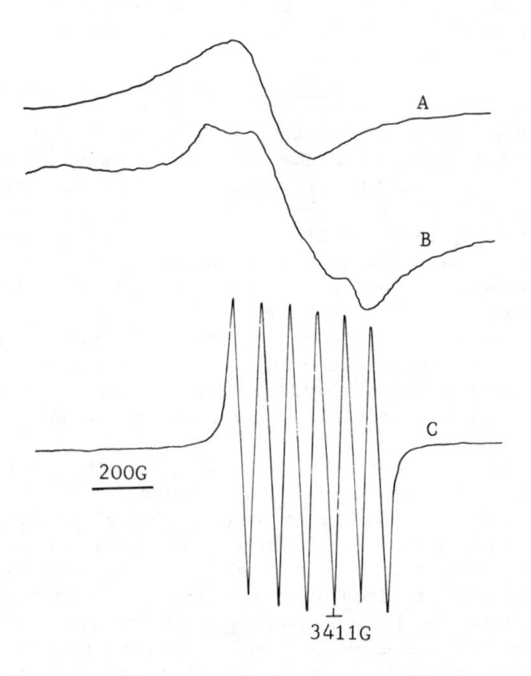

Figure 5. ESR spectra of Mn-SPS (2.65 mole %) A) neat, B) 5% solution in THF C) 5% solution in DMF.

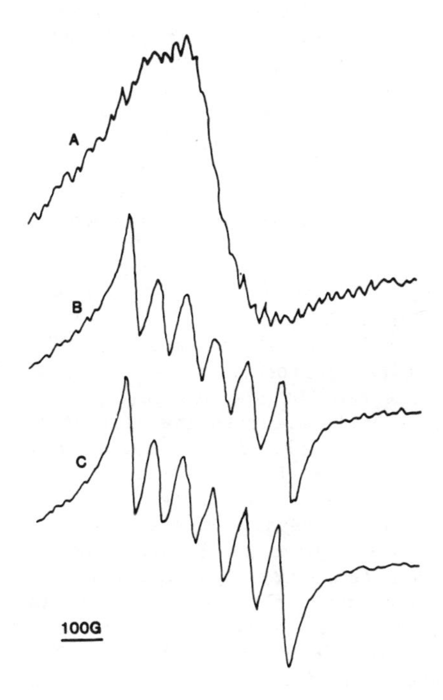

Figure 6. Effect of methanol cosolvent concentration on the ESR spectrum of Mn-SPS (1.15 mole%) in toluene/methanol solutions. A) 2% B) 5% C) 8%.

by Lundberg and co-workers(7,10) from viscosity data. At low alcohol concentrations, the Mn^{2+} ions are primarily associated, while at higher alcohol concentrations the population of isolated cations becomes significant.

As discussed in the introduction to this paper, different viscosity versus concentration behavior is observed for SPS solutions in toluene/methanol and in DMF. Polyelectrolyte behavior is observed only in the latter solvent. The ESR spectrum of a 2.65 mole % Mn-SPS in these two solvents was studied at various concentrations. For both solvents, the hyperfine structure characteristic of isolated Mn^{2+} ions was observed in very dilute solutions and at concentrations for which Lundberg and Phillips(10) observed strong intermolecular interactions. The ESR data indicated that in dilute solution in both DMF and toluene/methanol, the Mn^{2+} exists mainly as isolated cations. In addition, the IR spectra indicated that the cation is removed from the anion to a similar degree in both solvents. Yet, a polyelectrolyte effect is observed experimentally only in DMF solutions. Although there was some dipole-dipole broadening of the toluene/methanol spectrum, the line width and the g-factor (g $\sim$2,000) in both cases were identical. The g-factor of 2.000 is characteristic of an isolated Mn^{2+} in solution(21).

The ESR spectra of a 2.65 mole % Cu-SPS in bulk and in 5% solutions of 95% toluene/5% methanol and DMF are given in figure 7. In all cases, no signal was observed above 4000G. Similar results were obtained at higher sulfonate concentrations. Differences in the fine structure of the ESR spectra do, however, indicate differences in the Cu^{2+} environment in the different samples.

The ESR spectrum of isolated Cu^{2+} consists of four lines due to hyperfine interactions with the Cu^{2+} nuclear spins(26). Dipole-dipole and exchange interactions between electron spins of neighboring Cu^{2+} result in broadening of the ESR spectrum. The solid state spectrum of Cu-SPS given in Figure 7 shows a single broad line that may be due to interactions between Cu^{2+} ions. More detail is seen in the spectra of the Cu-SPS in solutions of toluene/-methanol and DMF, though the hyperfine structure is still not well resolved. Although additional study of the ESR spectra of Cu-SPS is warranted, these preliminary results are in essential agreement with the Mn-ESR results. That is, polar solvents are able to solvate interactions between the cations in these ionomers.

Conclusions

The vibrational modes of the sulfonate group in lightly sulfonated polystyrene are affected by changes in its local chemical environment. In dry materials, the counterion imposes a strong electrostatic field on the sulfonate anion, which polarizes the S-O dipole and shifts both the asymmetric and symmetric stretch vibrations to higher frequency. The addition of water or organic solvents varying in dielectric constant from 7 to 40 weaken the polarization by the counterion and shift the stretching vibrations to lower frequencies, approaching the vibrational frequencies of the degenerate state of the anion, 1200 cm^{-1} and 1034 cm^{-1}. The results may be

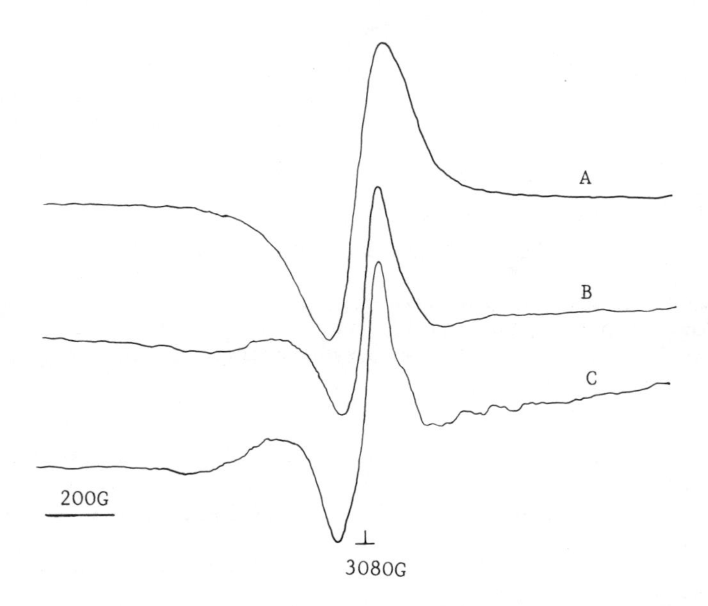

Figure 7. ESR spectra of Cu-SPS (2.65 mole %). A) neat, B) 5% solution in 95% toluene/5% methanol, C) 5% solution in DMF.

rationalized in terms of a solvation shell as proposed by Lowry and Mauritz(12) for Nafion ionomers.

Electron spin resonance spectroscopy was used to characterize interactions between the cations in bulk samples and in solution. In the solid state the Mn^{2+} cations of Mn-SPS are primarily associated. In solution the ratio between associated and isolated ions depends upon the polarity of the solvent. The more polar the solvent, the greater is the extent of isolation of the cations. Preliminary ESR spectra of Cu-SPS are in qualitative agreement with this conclusion.

Acknowledgments

We wish to thank Mr. Tae Ho Yoon who contributed several of the spectra and Profs. Harry Frank (Univ. Conn.) and Bruce Kowert (St. Louis Univ.) for their help in interpreting the ESR spectra.

The material is based upon work supported by the National Science Foundation under Grant No. DMR8407098 and the Petroleum Research Fund of the American Chemical Society under Grant No. PRF 15614-AC7.

Literature Cited

1. W. J. MacKnight and T. R. Earnest, Macromol. Rev., 16, 41 (1981).
2. C. G. Baziun and A. Eisenberg, Ind. Eng. Chem. Prod. Res. Dev., 20, 271 (1981).
3. W. J. MacKnight and R. D. Lundberg, Rubber Chem. Technol., 57(3), 652 (1984).
4. A. Eisenberg and M. King, "Ion-Containing Polymers", Academic Press, N.Y., 1977.
5. A. D. Wilson and H. J. Prosser, Eds., "Developments in Ionic Polymers -1.", Appl. Science Publ., London, 1983.
6. E. Otocka, J. Macromol. Sci.-Rev. Macromol. Chem., C5(2), 275 (1971).
7. R. D. Lundberg and H. S. Makowski, J. Polym. Sci., Polym. Phys. Ed., 18, 1821 (1980).
8. H. Schade and K. Gartner, Plaste Kautschuk, 21(11), 825 (1974).
9. C. Rockas, A. Domard, and M. Rinardo, Polymer, 20, 76 (1979).
10. R. D. Lundberg and R. R. Phillips, J. Polym. Sci., Polym. Phys. Ed., 20, 1143 (1982).
11. R. A. Weiss, S. R. Turner, and R. D. Lundberg, J. Polym. Sci., in press.
12. S. R. Lowry and K. A. Mauritz, J. Am. Chem. Soc., 102, Y665 (1980).
13. M. Falk, in "Perfluorinated Ionomer Membranes", A. Eisenberg and H. L. Yeager, Eds., American Chemical Society, Washington, DC, 1982, p.139.
14. R. A. Komoroski and K. A. Mauritz, J. Am. Chem. Soc., 100, 7487 (1980).
15. H. S. Makowski, R. D. Lundberg, and G. H. Singhal, U. S. Patent 3,870,841 (1975).

16. C. Y. Liang and S. Krimm, J. Polym. Sci., 27, 241 (1958).
17. R. N. Haszldine and S. M. Kidd, J. Chem. Soc., 290 (1955).
18. C. La Lau and E. A. M. F. Dahmen, Spectrochim Acta, 11, 594 (1957).
19. G. Zundel, "Hydration of Intermolecular Interaction", Academic Press, New York, 1969.
20. D. Rahrig, W. T. MacKnight, and R. W. Lenz, Macromolecules, 12(2), 195 (1979).
21. C. C. Hinckley and L. O. Morgan, J. Chem. Phys., 44(3), 898 (1966).
22. W. B. Lewis, M. Alei, Jr., and L. O. Morgan, J. Chem. Phys., 44(6), 2409 (1966).
23. B. R. McGarvey, J. Phys. Chem., 61, 1232 (1957).
24. H. Toriumi, R. A. Weiss, and H. A. Frank, Macromolecules, in press.
25. M. Pineri, C. Meyer, A. M. Levelut, and M. Lambert, J. Polym. Sci., Polym. Phys. Ed., 12, 115 (1974).
26. R. Vasquez, J. Avalos, F. Volino, M. Pineri, and D. Gallard, J. Appl. Polym. Sci., 28, 1093 (1983).

RECEIVED June 10, 1985

4

Spectroscopic and Thermal Studies of Ionic Interactions in Ionomers

V. D. Mattera, Jr., S. L. Peluso, A. T. Tsatsas, and W. M. Risen, Jr.

Department of Chemistry, Brown University, Providence, RI 02912

Ionic interactions in ionomers are reflected in their spectra both directly and indirectly. They indirectly influence the spectral features associated with the polymeric backbone and the pendant sites as well as some of the spectral characteristics of polyatomic cations. Studies of these spectral properties are extensive. Ionic interactions are probed more directly by observing the vibrations of the cations at their anionic sites. Ion–motion vibrational bands, which occur in the far–infrared spectrum, have been studied in PFSA (Nafion) (1), PEMA (2–3), PSMA (4–5), and PSSA (6) ionomers with a range of cations, ionic site concentrations, and other conditions. The force field elements that can be derived from them reflect how the interionic forces vary with the nature of the ionomer.

The leading term of the attractive part of these forces is Coulombic in origin. The study of the Coulombic attraction of cations and anions in ionomers is important not only for understanding ion–transport and salting–in processes of sparingly soluble salts, but also for understanding the influence of ionic forces on microstructure or morphology. The key idea is that the interionic forces, together with the weak attraction between hydrophobic polymer segments, and the forces between hydrophobic and ionic components drive a process against entropic factors to organize the ionomers into ionic domains and hydrophobic microphases that have the minimum free energy attainable on the time scale of the experiments. This general idea has been discussed extensively and treated theoretically since its introduction and early development by Eisenberg (7,8), Ward and Tobolsky (9), MacKnight *et al* (10), and others. The main driving force typically is thought to be the interionic attraction. However, the balance of enthalpic and entropic effects that leads to having the free energy minimum occur for the microphase–separated or domain structure is rather delicate. The interionic forces must be strong enough to provide most of the energy minimization required to overcome disordering tendencies, but they must not be strong enough to impose a different order.

Of the microphase–structure dependent physical properties of ionomers, perhaps the most widely studied are glass transition temperatures, (Tg), and dynamic mechanical response. The contribution of the Coulombic forces acting at the ionic sites to the cohesive forces of a number of ionomeric materials has been treated by Eisenberg and coworkers (7). In cases in which the interionic cohesive force must be overcome in order for the cooperative relaxation to occur at Tg, this temperature varies with the magnitude of the force. For materials in which other relaxations are forced to occur at Tg, the correlation is less direct.

In this paper the use of far–infrared spectroscopy to measure cation–motion bands is presented by considering series of ionomers in which the natures of the cation and

0097–6156/86/0302–0054$06.00/0

anionic sites are varied. The role of cations in influencing the glass transition temperatures of one series of ionomers is considered by presenting the measured Tg values for sulfonated linear polystyrenes neutralized by alkali and alkaline earth cations.

Ion-Site Vibrations in Ionomers

The far-infrared spectra of dried alkali metal ionomers formed from perfluorocarbonsulfonic acid (PFSA) are shown in Figure 1, and those of analogous alkaline earth ionomers are shown in Figure 2. The PFSA studied was of Nafion type, had an equivalent weight of 1100 amu, and was 5 mil (1.3×10^{-4}m) thick. In each spectrum there is a strong sharp band near 200 cm^{-1} which is due to an internal vibrational mode of the PFSA and a major band that is characteristic of the cation present. As shown in Figure 3a, the frequencies of these bands in the alkali PFSA ionomers decrease as the cation mass increases. The same trend is observed for the alkaline earth materials. The dotted connecting lines join cations of approximately the same mass, (eg. Sr^{+2} and Rb^{+}), and show that for essentially the same mass the cation-motion frequency increases as the charge on the cation increases. Clearly, the frequency of each feature depends strongly on the mass and charge of the cation present. The bands are assigned to cation-motion vibrations, which means that the vibrational motion involves the motion of the cation and its site. Although the reduced mass of the vibration must involve more than the cation, using the lowest order approximation, in which the cation oscillates against a very heavy surrounding, one can see a cation-mass dependence of the band frequencies by plotting $M^{-1/2}$ versus ν, as shown in Figure 3b.

The variation of the cation-motion frequency with cation charge shows that the spectrum probes a force field in which Coulombic interactions are important. This leads to the conclusion that it also should vary significantly with the charge on the anionic site.

Three types of variation of the anionic site from that of the PFSA materials are useful to consider. One involves changing the polymeric backbone to which the sulfonate, SO_3, units are attached. This should cause the effective charge at the site to vary. The second involves changing the site altogether, for example, to CO_2^- sites, as found in PSMA (polystyrenemethacrylic acid) or PEMA (polyethylenemethacrylic acid) ionomers. And the third, which is more difficult to achieve with certainty, involves varying the microstructure of the ionic region.

The class of sulfonated linear polystyrene ionomers prepared by Lundberg and coworkers (11) at Exxon have sulfonate groups on aromatic rings that are pendant from the polymer backbone. This electronic environment should withdraw electrons from the atoms of the SO_3^- site less than the fluoroether chains to which they are attached in the PFSA ionomers. Thus, the effective negative charge of the anionic sites in PSSA ionomers should be higher. The far-infrared spectra of dry alkali metal PSSA films, formed from PSSA materials in which 6.9 mole percent of the rings have been sulfonated, are shown in Figure 4. As can be seen in Figure 5, they and analogous M^{+2}-PSSA films have cation motion bands that shift with cation mass in the manner expected from the lowest order vibrational approximation. The frequencies in the PSSA materials are uniformly higher than those of the analogous PFSA ionomers.

Ionomers with carboxylate anionic groups exhibit both the effect of changing the nature of the anionic site and the effect of microstructural changes. Polyethylenemethacrylic acid (PEMA) ionomers, such as the Surlyn materials made by DuPont, have this anionic group. The positions of the cation motion bands of the alkali metal PEMA ionomers have been found to be even higher than those for the analogous M^{+}-PSMA ionomers. They are listed in Table 1, where they can be compared to the cation-motion bands of the M^{+}-PSSA and M^{+}PFSA ionomers.

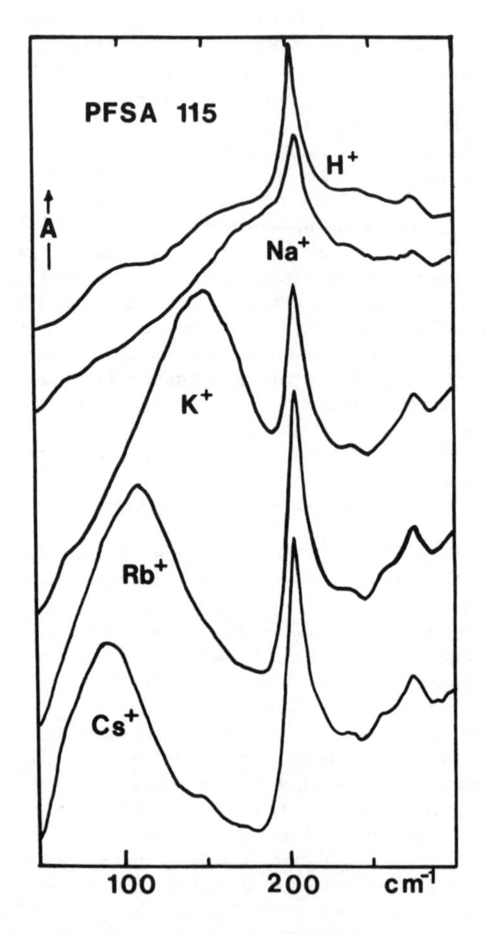

Figure 1. Far infrared spectra of dry MPFSA (M = Na, K, Rb, Cs) films of equivalent weight 1100 and thickness 1.3 x 10^{-4} m.

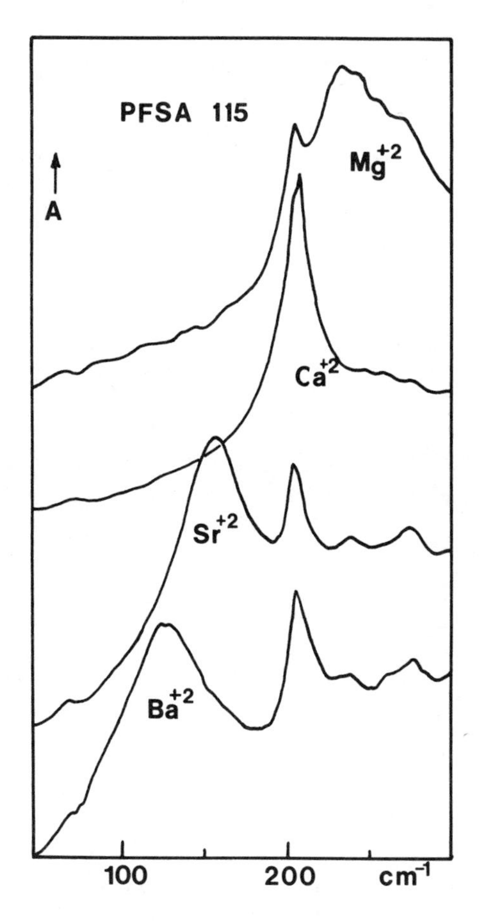

Figure 2. Far infrared spectra of dry MPFSA (M = Mg, Ca, Sr, Ba) films of equivalent weight 1100 and thickness 1.3×10^{-4} m.

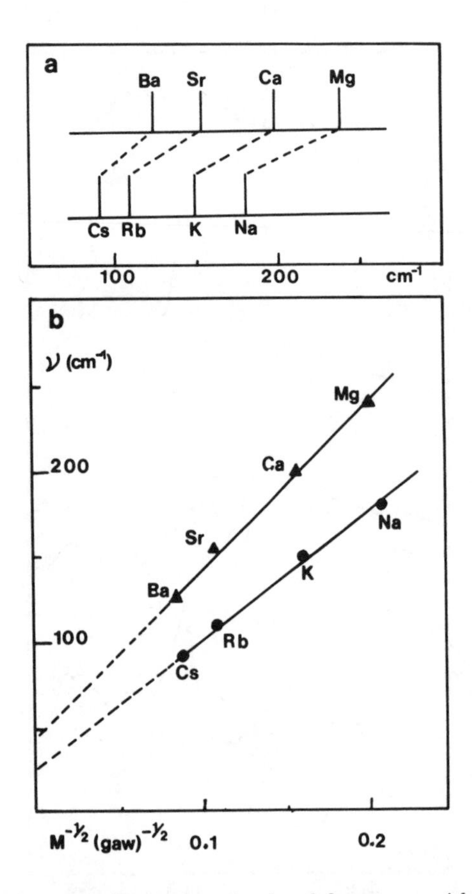

Figure 3. Variation of cation–motion band frequency with cation: (a) comparison of frequencies for cations of approximately the same mass but different charge; (b) plot of frequency versus inverse square root of cation mass.

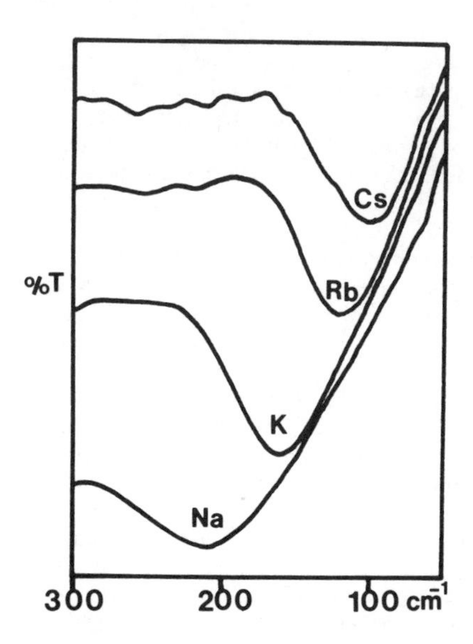

Figure 4. Far infrared spectra of dry MPSSA (6.9) (M = Na, K, Rb, Cs) films.

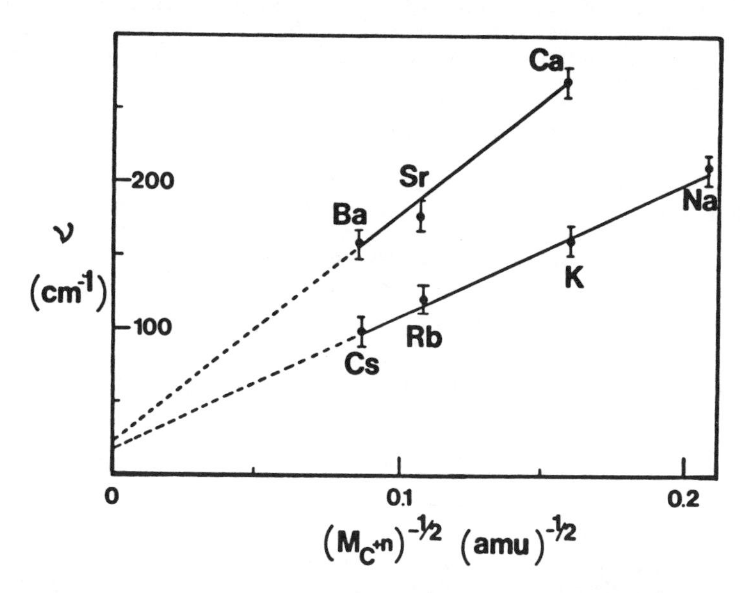

Figure 5. Variation of cation–motion band frequency in MPSSA (6.9) ionomers with inverse square root of cation mass. Reproduced with permission from Ref. 6. Copyright 1984, John Wiley & Sons, Inc.

When the structure of the cation–containing environment changes significantly, the frequency of the vibration involving the cation is expected to change as well. Development of large aggregations of anions and cations in ionomers can give rise to a lower frequency because the reduced mass of the vibration increases and the force field decreases as the interionic Coulombic interactions are screened. This was observed in the case of sodium PSMA ionomers. As the concentration of anionic sites was increased in that case, the fraction of the cations in large clusters increased and a new cation–motion band appeared at a lower frequency than that of the band for similar polymers with lower methacrylate concentration.

For each of the ionomer systems studies as well as for cations in zeolites, the cation–motion bands are easy to identify, because they vary in frequency approximately as $M^{-1/2}$, where M is the mass of the cation. However, $M^{-1/2}$ clearly is not the correct complete reduced mass for the vibration, since that would require the site to be of infinite mass. A better approximation to the proper reduced mass, μ, can be obtained by considering at least the atoms immediately surrounding the cation. Typically, μ is calculated from models in which the cation has either an octahedral surrounding (for the T_{1U} mode) or a tetrahedral surrounding (for the T_2 mode). With it and the band frequency, the force field element, (force constant), for the vibration can be calculated.

Since the vibrational frequencies reflect both the nature of the cation and the anion and, more specifically, vary with the cationic charge and the effective anionic charge, it is of interest to try to put this information on a more quantitative basis. A potential energy function describing this type of interaction can be constructed in the form of a modified Rittner potential (3). The Rittner potential contains the well known ion–multipole terms of classical electrostatics, the van der Waals term, and a Born–Mayer repulsion term. The degree to which the cation motion bands reflect actual ionic interactions can be investigated readily by taking the two main terms, the Coulombic attraction and the repulsion potential, and computing the dependence of the frequency on the properties of the cation and the effective charge, q_A, of the anionic site. This lead term functional form, which includes the terms that vary significantly with vibrational amplitude, is

$$U = -q_A q_C e^2/r + \exp(-r/\rho)$$

where q_C is the value of the cation charge, r is the cation–oxygen separation, and ρ is a constant equal to 0.33A. Setting its first derivative equal to zero, its second derivative can be evaluated at the value of r corresponding to the minimum. This is the force constant, F, which is related to the bond frequency, $\nu\,(\mathrm{cm}^{-1})$ by $4\pi^2 c^2 \nu^2 = F/\mu$. Thus,

$$\nu = [q_A e^2/(2\pi c)^2]^{1/2} [q_C (r - 2\rho)/\mu r^3 \pi]^{1/2}$$

which can be written as a linear equation $\nu = mX + b$. Here, X contains the cation-dependent terms, and the slope m scales as the square root of the effective anionic charge.

A plot of ν versus X for PSSA ionomers, Figure 6, yields a reasonably straight line, which indicates that the use of an essentially Coulombic attractive potential describes the interaction over this range of mass and charge variation. As shown in Figure 7, the cation–motion bands for both the MPEMA and MPFSA ionomers also fall on reasonably straight lines. The higher slope for the MPEMA system may indicate that the carboxylate groups have a higher effective charge. Perhaps a more meaningful comparison is between the two classes of sulfonate–containing ionomers in which the local geometries and groups of atoms are the same. The value of m for MPSSA ionomers is 970 $\mathrm{cm}^{-1}\,\mathrm{amu}^{1/2}\,\mathrm{A}^{3/2}$, while that for the MPFSA ionomers is lower at 642 $\mathrm{cm}^{-1}\,\mathrm{amu}^{1/2}\,\mathrm{A}^{3/2}$.

Table 1. Selected Cation Vibrational Frequencies (cm^{-1})

Cation (M^+)	PEMA ($-CO_2^-$)	PSSA ($-SO_3^-$)	Zeolite (SiteII)	PFSA ($-SO_3^-$)	Zeolite (SiteI)
Na^+	230–250	210	180–190	180	160–170
K^+	180	160	133–155	150	107
Rb^+	-	120	-	110	-
Cs^+	115–135	100	-	90	62–86

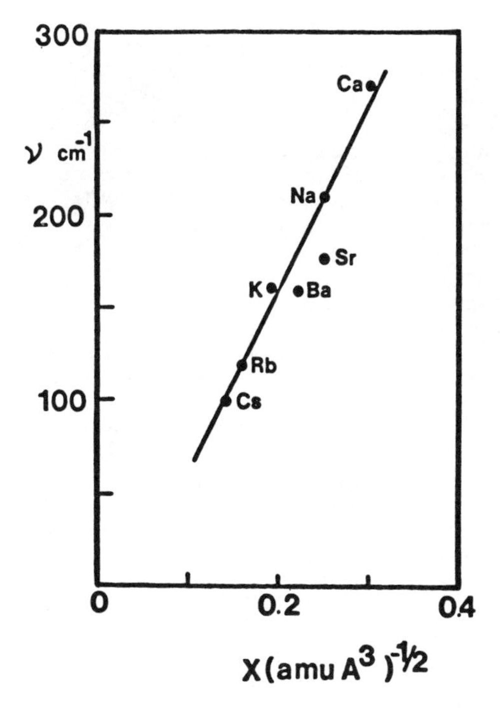

Figure 6. Plot of X (defined in text) versus cation motion frequency for MPSSA (6.9) ionomer films.

Cation Dependence of Tg in PSSA Ionomers

The glass transition temperatures of ionomers have been widely studied and the contribution of a number of compositional variables have been considered. However, it remains of interest to investigate the variation of Tg as the cation and the anionic concentration are varied in a series of ionomers of the same type. To do this, the values of Tg were obtained by differential scanning calorimetric (DSC) methods on Na, K, Rb, Cs, Mg, Ca, Sr and Ba PSSA materials with 3.4, 6.9, 12.7, and 16.7 mole percent sulfonation (y). They were measured on materials that had been completely ion–exchanged with salt solutions in methanol or ethanol and then dried with heating under vacuum. The DSC scans were measured once on heating the materials from 20 to 200°C, then once after quenching from 200 to 20°C, and finally once after annealing by slow cooling from 200°C. The Tg values obtained on heating the samples that had been quenched are used for comparison and are given in Table 2.

The averages of the Tg values for the alkali metal and alkaline earth forms of each of the degrees of sulfonation are included in Table 2. Two trends are evident. First, among the M^{+} ionomers Tg depends largely on the degree of sulfonation rather than on the detailed properties of the cation. Second, among the M^{+2} ionomers there is little dependence on either the cation or the degree of sulfonation up to at least 12.7 mole percent. At y = 16.7, each of the cation PSSA series has a value of Tg of about 126°C.

The PSSA ionomers have attracted considerable interest since they were reported by Lundberg and coworkers (11,15). At issue is the question of whether ionic domains form, what structures exist in them, and how these depend on ionic concentration, cation and thermal history. On the basis of the effect on the viscoelastic properties of PSSA ionomers caused by varying the metal sulfonate composition, Rigdahl and Eisenberg (16) concluded that the onset of ion clustering occurs at about 6 mole percent. This is supported by the fluorescence studies by Okamoto, *et al* (17), which indicated that no appreciable ion clustering occurs below 6 mole percent. However, Cooper and coworkers (18) reported that the results of their EXAFS experiments indicate that ion clustering is present at about 3 mole percent in ZnPSSA ionomers. Lundberg and associates (15) reported that their SAXS experiments indicate that clusters are present at compositions as low as 3 mole percent, resulting in a microphase separated material. The small angle neutron scattering (SANS) study of PSSA ionomers reported by MacKnight and coworkers (19) suggests the presence of ionic clusters. And, recently Weiss and coworkers (20,21) have reported several thermal, SAXS and ESR studies of several PSSA ionomers which are interpreted in terms of ionic clustering.

In many of these studies the postulated formation of ionic domains follows the patterns generally expected from the idea that domain formation is favored by increased concentration of anionic sites and strengthened anion–cation interaction force. However, Weiss *et al* found that the behaviors of the Na- and Zn-PSSA materials were not as expected. Thus, they found that the annealing of compression molded NaPSSA tended to increase the SAXS–measured distance associated with ionic interactions and that the NaPSSA (5.5% sulfonate) did not flow when heated at 300°C. On the other hand, annealing ZnPSSA (5.5% sulfonate) did not increase this distance, in fact it decreased it on the average while leaving the main peak unchanged, and the material did flow at 300°C. These observations made it appear that the ionic interactions with Na^{+} counterions were stronger than with Zn^{+2} counterions.

The Tg data helps in understanding the varied behavior of the PSSA ionomers. They show that up to about 13 mole percent sulfonation, that is when there are 8 or more styrene units between anionic sites, the Tg's of alkaline earth PSSA's do not vary strongly with either the nature of the cation or with y. The Tg's of alkali metal PSSA's in this composition range do vary with y but they do not vary systematically

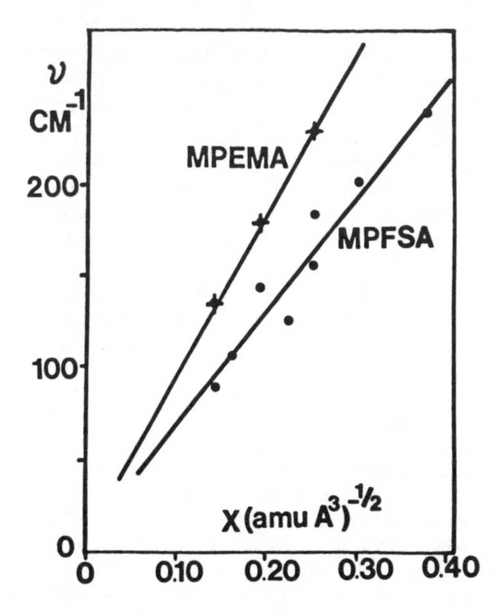

Figure 7. Plot of X (defined in text) versus cation motion frequency for MPEMA and MPFSA ionomer films.

Table 2. Tg(C) of M-PSSA Ionomers (y% SO_3)

M	y=3.4	y=6.9	y=12.7	y=16.7
Na	106	114	126	125
K	102	118	128	126
Rb	108	120	125	125
Cs	103	109	127	127
Mg	107	109	110	122
Ca	100	104	110	127
Sr	101	107	102	128
Ba	107	114	105	—

with cation. The data also show that when the sulfonation is so high (16.7%) that there are 6 or fewer styrene units between anionic sites and all are within 1 or 3 units of such a site, the Tg's of both the M^{+} and M^{+2}-PSSA's are about 126°C. Both of these types of cations must associate with the anionic sites on PSSA. As expected and shown by the values of the cation-site force constants (6), the M^{+2} cations associate with them more strongly. However, it is the M^{+} cations that have the more pronounced effect on Tg. This suggests that the roles of the cation-anionic site interactions in M^{+}-PSSA and M^{+2}-PSSA are somewhat different.

On the basis of these data and as suggested earlier, it is postulated that ionic domain formation is important in the M^{+}-PSSA ionomers, but that in the case of these particular M^{+2}-PSSA ionomers, at least, the ionic interactions are more closely modeled by random intermolecular and intramolecular crosslinks. The interaction of +2 cations with two anionic sites is quite strong, and there should be little or no free energy advantage to formation of larger clusters or domains. There clearly is no entropic advantage, since organization of both putative ionic domains and hydrophobic domains is disfavored entropically. And, it is reasonable to assume that the formation of an A^{-}-M^{+2}-A^{-} structure, which completes the cation's coordination, is not only energetically stable but also leaves the system with the greatest freedom to optimize the hydrophobic interactions. The randomness and independence of these crosslinks leaves many of the chain segments between them without any additional constraints from ionic interactions. Thus, up to about 13 mole % sulfonation (8 units between sites), most chain segments are nearly as free to be activated collectively at the glass transition as they would be if the ionic interactions were not present. This means that Tg will not depend strongly on either the nature of the +2 cation or the degree of sulfonation in this composition range. It also means that the material can flow at higher temperatures without the average cation-site distance changing, and may help explain the observation to this effect for ZnPSSA.

The formation of ionic domains in the M^{+}-PSSA materials has a different effect on Tg. The geometric requirements imposed by the domain formation that optimizes the anion-cation interactions cause a larger fraction of the chain segments between anionic sites to be strained than would be strained if there were only random crosslinks. This fraction is high enough that the cooperative activation of the styrene units between sulfonation sites occurs at increasing temperatures as the degree of sulfonation increases. Thus, Tg rises with increasing sulfonation. The Tg does not depend strongly on the nature of the M^{+} cation since all of them find their lowest available free energy structure to be domains. The M^{+}-PSSA materials that are formed in a random manner, such as rapid precipitation, can be expected to form more extensive domains and less viscous materials upon heating. This may explain the fact that Na-PSSA (5.5) did not flow when heated but it did increase the SAXS-measured distance (21).

In the case of both M^{+2} and M^{+} PSSA ionomers it is to be expected that over long periods of time the anion-cation interactions will approach their optimum (local) configuration by making small distance and angle changes at the expense of the weaker orientational forces of the neighboring styrene units. The temperature at which the additional strain of these units will be relieved is, of course, lower than that at which cooperative activation of styrene chain segments occurs. For polystyrene itself the latter occurs at *ca* 100°C, so it is reasonable to assign the LTE (low temperature endotherm) observed at 70°C for the PSSA ionomers to the relaxation of these local strains. This process is essentially the same as that described as densification of polymeric glasses during ageing.

Acknowledgments

The partial support of this work and the use of the facilities of the Materials Research Laboratory and the support of the Office of Naval Research are gratefully acknowledged.

Literature Cited

1. S. L. Peluso, A. T. Tsatsas and W. M. Risen, Jr., ONR Report TR 79-01, to be published, S. L. Peluso, Ph.D. Thesis, Brown University, 1980.
2. A. T. Tsatsas and W. M. Risen, Jr., *Chem. Phys. Lett.* 1970, **7**, 354.
3. A. T. Tsatsas, J. W. Reed and W. M. Risen, Jr., *J. Chem. Phys.* 1971, **55**, 3260.
4. G. B. Rouse, A. T. Tsatsas, A. Eisenberg and W. M. Risen, Jr., *J. Polym. Sci. (Polym. Phys. Ed.)*, 1979, **17**, 81.
5. G. B. Rouse, A. T. Tsatsas and W. M. Risen, Jr., *Chemika Chron.* 1979, **8**, 45.
6. V. D. Mattera, Jr. and W. M. Risen, Jr., *J. Polym. Sci. (Polym. Phys. Ed.)* 1984, **22**, 67.
7. A. Eisenberg, *Macromolecules* 1970, **3**, 147, 1971, **4**, 125.
8. A. Eisenberg and M. King, Eds., *Ion Containing Polymers, Vol 2 Polymer Physics*, Academic Press, New York, 1977.
9. T. C. Ward and A. V. Tobolsky, *J. Appl. Polym. Sci*, 1967, **11**, 2403.
10. W. J. MacKnight, W. P. Taggart and R. S. Stein, *J. Polym. Sci. Polym. Symp.* 1974, **45**, 113.
11. W. Siebourg, R. D. Lundberg and R. W. Lenz, *J. Polym. Sci. Polym. Phys. Ed.* 1980, **13**, 1013.
12. R. O. Lundberg and H. S. Makowski, *J. Polym. Sci. Polym. Phys. Ed.* 1980, **18**, 1821.
13. R. D. Lundberg and R. R. Phillips, *J. Polym. Sci. Polym. Phys. Ed.* 1982, **20**, 1143.
14. H. S. Makowski, R. D. Lundberg and G. H. Singbal, U. S. Patent 3, 1975, 870, 841.
15. D. G. Peiffer, R. A. Weiss and R. D. Lundberg, *J. Polym. Sci. Polym. Phys. Ed.* 1982 **20**, 1503.
16. M. Rigdahl and A. Eisenberg, *J. Polym. Sci. Polym. Phys. Ed.* 1982, **20**, 1143.
17. Y. Okamoto, Y. Veba, N. F. Ozanebakov, and F. Bands, *Macromolecules* 1981, **14**, 17.
18. D. J. Yarusso, S. L. Cooper, G. S. Knapp and P. Georgopoulos, *J. Polym. Sci. Polym. Lett. Ed.* 1980, **18**, 557.
19. T. R. Earnest, Jr., J. S. Higgins, D. L. Handlin and W. J. MacKnight, *Macromolecules* 1981, **14**, 192.
20. R. A. Weiss, *J. Polym. Sci. Polym. Phys. Ed.* 1982, **20**, 65.
21. R. A. Weiss, J. Lefalar and H. Toriumi, *J. Polym. Sci. Polym. Lett. Ed.* 1982, **20**, 661.

RECEIVED December 13, 1985

5

Chemistry in Ionomers

D. M. Barnes, S. N. Chaudhuri, G. D. Chryssikos, V. D. Mattera, Jr., S. L. Peluso, I. W. Shim, A. T. Tsatsas, and W. M. Risen, Jr.

Department of Chemistry, Brown University, Providence, RI 02912

Ionomers are polymers that are functionalized with ionic groups (usually anionic sites) attached at various points along polymeric backbones that are not extensively crosslinked (1–2). Such materials have a tendency to form ionic domains in which the anionic groups and their associated cations are microphase separated from the typically hydrophobic portions of the polymer. Thus, the ionic domains formed are isolated by a medium of low dielectric constant (i.e. the polymeric backbone) although, in some cases, hydrophilic channels have been reported to connect adjacent ionic domains (3). The size and structures of these domains vary with the nature of the cation, the stoichiometry of the polymer, the degree of solvation of the system and the method of preparation. They can be as small as ion-pairs or small multiplets, but in some cases they have been reported to be in the 20–100 Å diameter range.

Transition metals ions can be incorporated in ionomers by a variety of methods. One involves the neutralization of the acid form of the ionomer through formal acid-base reactions. This occurs, for example, when the acid form of sulfonated polystyrene is reacted in solution with a metal acetate, or when an alkali hydroxide is reacted directly with molten polyethylene methacrylic acid. The morphologies of such ionomers depend on how they are treated, and, in particular, on whether they are annealed or desolvated. Another method involves ion exchange of dissolved transition metal cations for alkali or hydrogen ions in insoluble ionomer films. In this case, the domain structure possessed by the original ionomers can be retained, because the ion-exchange process need not disrupt or reorganize the hydrophobic regions significantly. Thus, it is possible to obtain a material that has domains containing transition metal ions.

These transition metal containing domains can be thought of as small chemical reactors since they can potentially confine reactant molecules in the proximity of metal ions. Some reactants (such as solvent molecules or H^+) can already exist in the domains after ion exchange treatment. Others, such as CO, H_2, NO, C_2H_2, C_2H_4, N_2H_4, incorporated after appropriate exposures of the membranes, can be confined for long enough time for the species in the domains to interact with them as they would if they were surface sites exposed to gaseous reactants at much higher pressures. This is to say that the effective reaction time and effective concentration (or pressure) can be higher, and reactions can occur that otherwise would require more extreme conditions. Moreover, if the metal ions in the domains can be treated to create active sites, potential catalytic centers could be produced.

The ionomers have other potential chemical applications as well. For example, upon reduction of the metal cations, metal particles could be formed with a size distribution controlled by the domain-morphology of the membrane. If the reducing

0097-6156/86/0302-0066$06.00/0

agent is H_2, H^+ will be provided adjacent to the metallic species within the isolated domain-reactor.

The characteristics discussed above are mainly related to clustering in the ionic phase, but the role of the hydrophobic phase also is quite important. In some cases it controls the gas transport properties of the material (e.g. O_2 through PFSA) (4). And, it makes it possible to keep hydrophobic reactions in the neighborhood of the ionic domain species (5). Moreover, metal complexes with bulky hydrophobic ligands can be supported in the ionomers because of synergystic interaction of both polymer phases (6). Interesting electrocatalytic or photocatalytic systems take advantage of these unique properties of ionomers (7-8). Moreover, support of the reactants in ionomers may be useful for reactant/product separations.

The studies described in this paper were designed to use the special properties of ionomers, but they are not the only type of chemistry that does so. Other chemical systems that employ ionomer properties include their use as sources of protons of high effective acidity (superacidity) in the catalysis of organic reactions (9-11) and their use as integral components of chemically active membranes (e.g. cell dividers or electrode coatings) (12).

Ionomers are certainly not the only useful support for transition metals and ions. Indeed, inorganic oxides, such as silica, zeolites and aluminas, are the most widely used at present (13). Among the organic polymeric supports now used, the most closely related to the ionomers are the well known ion-exchange resins. While they are polyelectrolytic, as are the PFSA and PSSA ionomers, they are not thought to possess the potentially useful morphological properties of ionomers.

The study of chemistry in ionomers will be illustrated in three ways. The first involves several feasibility studies, specifically the reactions of metal ions with simple gaseous molecules. The second involves results of studies of systems of potential catalytic importance. And, the third illustrates the study of particle formation and reaction kinetics. They are illustrative of a body of work in this area (14-18).

Experimental Approach

The studies mentioned here were carried out on two different ionomers, PFSA (Nafion[R]) a perfluorocarbon polymer with sulfonic groups connected to the backbone by fluorether chains, and PSSA (sulfonated polystyrene), kindly provided by the Plastics Department of the E. I. DuPont de Nemours Corporation and Dr. R. D. Lundberg of the Exxon Research and Engineering Company, respectively. Since not all of the materials used were commercial products, they will be referred to here by their acronyms. The PFSA materials are similar or the same as Nafion and were used in the 900-1600 amu equivalent weight range. The films used were typically about 25 μm thick, which is convenient for infrared and uv-visible studies. The PSSA materials are sulfonated linear polystyrene and have formulae of the following form. Here, $a = y/(x+y)$ represents the mole fraction of sulfonate groups after the figure. (The materials studied have x values in the 0.03-0.21 range, and were formed into films of *ca* 25 μm thickness by solution casting.)

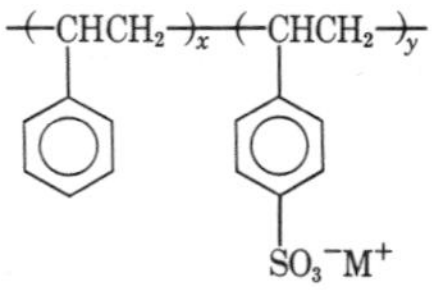

$$-(CHCH_2)_x-(CHCH_2)_y-$$

$$SO_3^-M^+$$

The transition metal ion-containing films were prepared carefully by ion-exchange as described elsewhere (10-16). These films were mounted in appropriate stainless steel or pyrex glass reactors/spectroscopic cells. Both reactors, described in detail elsewhere, had infrared (KBr) or uv-vis (quartz) transmitting windows for spectroscopic studies. They were equipped with vacuum valves for evacuation and gas admission, a heating system and a temperature monitor.

The gases used in these studies were of high purity (99.99% CO, Matheson Corp; 99.995% H_2, Airco, Inc.; 99.995% D_2, Matheson Corp., 99.5% O_2, Airco, Inc.; 99% ^{13}CO, Monsanto Research Corp; 99.0% NO, Matheson Corp; and 99.98% C_2H_4, Matheson Corp.). The NO and, as necessary, other gases were passed through purifying traps before use.

The infrared spectra of the ionomer films were measured using a Digilab 15B FT-IR Spectrometer. Spectra typically were measured in the 3800–400 cm^{-1} region with 2 cm^{-1} resolution by Fourier transforming 400 interferrograms, although many spectra were measured over other narrower ranges at higher resolution, as appropriate. The uv–vis spectra in the 700–300 nm region were recorded with a Cary 17 uv–vis Spectrophotometer at a typical resolution of 1 nm.

Electron microscopic measurements were made on film samples embedded in Spurr's low viscosity medium and microtomed into sections which were thin supported on Cu grids. Measurements were made on a Phillips Model 401 LS TEM operated at 100 KV in the bright field mode at calibrated magnifications up to 500,000x. At this magnification the instrument is capable of resolving features that are about 3.5 Å in diameter.

Feasibility of Approach: Reactions of Ag, Cu, Fe, Ni, and Co with CO and NO

In order to test the feasibility of the notion that transition metals in ionomers would exhibit interesting reactivity, several important types of metal ions in PFSA ionomers were reacted with CO and NO under mild conditions. The ions Ag^{+} and Cu^{+2} were chosen because their carbonyls are known to be quite difficult to form without chemically forcing conditions on high surface area supports. The ions Fe^{+3}, Ni^{+2} and Co^{+2} were chosen because, although their elements have well known carbonyls (e.g. $Fe(CO)_5$, $Ni(CO)_4$, and $Co_2(CO)_8$), the carbonyls typically are prepared by reaction with the reduced forms of the metal. In fact, $Ni(CO)_4$ forms spontaneously by reacting metallic Ni with CO at low pressure. On the other hand, most of these ions form nitrosyls by reaction with NO under mild conditions.

As shown in Figure 1, when Ag(I)–PFSA was placed in an evacuated reactor it reacted with CO at low pressure (*ca* 0.1 atm) to form a silver carbonyl, which absorbs light at 2178 cm^{-1}. This reaction occurs whether the AgPFSA film is used as prepared by ion–exchange or whether it is dehydrated before its exposure to CO (14). The C–O vibrational frequency indicates that the Ag–CO bonding is essentially sigma in character, since association between a σ^* CO orbital and Ag^{+} without accompanying M–π^* interaction should increase the CO frequency from the value of *ca* 2143 cm^{-1} it would have if it were essentially unperturbed. The formation of a silver carbonyl in significant amount under such mild conditions is quite unusual.

On the other hand, although the formation of carbonyls with Fe, Ni and Co is very common, it does not occur when PFSA films exchanged with their aquated Fe^{+3}, Ni^{+2}, Co^{+3}, Co^{+2}, or Cu^{+2} ions are reacted, either hydrated or dehydrated, with CO in the 0.1 to 1.0 atm and 25°C to 220°C range (15). It seems clear that reduction of these ions is required prior to carbonyl formation. While that fact is well known in metal carbonyl synthesis, it isn't entirely clear why such reduction is necessary in this ionomer since a wide range of other metal ions in PFSA do form carbonyls. In some of these cases it may be that the hydration of the ions is too strong for the CO to displace H_2O to form a relatively weak association. In other cases the reduction may simply require more of a driving force than can be provided by the H_2, which is formally or actually formed in the films through reaction of CO and H_2O.

Other feasibility studies have focused on the study of relatively well known types of reaction. Thus, for example, $Cu(NH_3)_4^{+2}$ was found by exposing Cu(II)PFSA to NH_3(g). A reaction which served to put this system into its general chemical context was that of dry Cu(II)PFSA with NO(g) (14). As shown in Fig 1, a copper nitrosyl is formed which exhibits a ν(NO) of 1924 cm^{-1}. Other important reactions with NO(g) have been carried out with Rh (15), Ru (15), Pd (14), Co (15), Ni (15), and Fe (15).

In addition to the simple addition reactions with CO, NO and NH_3, a range of other reactions between transition metals in PFSA and PSSA ionomers and reactive species have been carried out. These include the following systems: Pd–PFSA with CO, NO, and C_2H_4 (14); Rh–PFSA with H_2, C_2H_4, CO, C_2H_2, and N_2H_4 (15,17); Ru–PFSA with C_2H_2, N_2H_4, O_2, CO and H_2 (15,17); Pt–PFSA with H_2, N_2H_4, CO, and C_2H_2 (16,17); Ir–PFSA with H_2, CO, and C_2H_2 (16); Ag–PFSA with H_2 and O_2 (20); Ru–PSSA with CO, H_2, O_2, and ROH (18,21); and Rh–PSSA with CO, H_2 and H_2O (17,19).

Reactions of Potential Catalytic Interest: Transition Metal Ionomers

A number of the reactions of metals in ionomers are of catalytic interest for three reasons. The first, obviously, is that catalytic reactions may occur and may proceed faster or under milder conditions than they do with other supports. The notion is that this could happen because the ionic domain could confine reactant molecules in the region of the metallic species for a relatively long residence time and, in some cases, because the acid anions in the domains could assist the reaction. The second main reason is that the domain structure could impose a morphological constraint on the reactions. The imposition of this constraint on the process of reduction of metal ions to metallic particles, for example, could lead to forming a material with small particles with a narrow particle size distribution. Finally, the third reason is that the ionomers can effect separations through either their ion exchange or their gas diffusion characteristics. These separations could be good enough to make the ionomers useful catalysts even if the rates of the catalyzed reactions were not the highest attainable ones.

As indicated in the previous section, a range of reactions of transition metal ionomers of potential catalytic interest have been studied (14–23). While space does not permit presenting the results in detail here, it is appropriate to illustrate several of the types of results that have been obtained. The first involves reactions that can be compared readily to reactions of the same metals or ions on other supports. The second type demonstrates the formation of metal particles by reduction of metal ionomers. And, the third type concerns the catalytic potential of these types of systems.

Among the most useful transition metal catalysts are ones containing supported Ru or Rh. Reactions of supported Rh with CO have been very widely studied. The reaction of Rh–PFSA materials with CO yields carbonyl species that are quite similar to those observed on supported or single crystal Rh (15,17). The ν(CO) region of the infrared spectrum of partially dehydrated Rh(III)PFSA and of this material after exposure to CO at pressures from 0.1 to 0.5 atm, followed by evacuation of the infrared cell, are shown in Fig 2. The bands in these spectra are due to carbonyls formed at Rh sites of a range of oxidation state and molecularity. For example, the high frequency band, at 2165 cm^{-1}, which disappears on evacuation and on subsequent reaction is due to CO weakly adsorbed on Rh(III), and those at *ca* 2110 and 2050 cm^{-1} are due to a Rh(I) *cis* dicarbonyl. The reduction to the latter form is thought to occur as a result of the formation, formally or actually, of H_2 from the CO + H_2O reaction in the domains. When Rh(III)PFSA is reduced directly with H_2, it reacts with CO to form species assigned as $Rh(I)(CO)_2$ (2110 and 2050 cm^{-1}), Rh(0)(CO) (2080–2060 cm^{-1}), and Rh(0)μ–(CO) (1830–1890 cm^{-1}). These are quite similar to those seen on such supported, reduced Rh species as $Rh(Al_2O_3)$ (24).

The reactions of RuPSSA are even more enlightening, because they not only show the species found in earlier work on supported Ru but they also have permitted observation of a widely postulated hydride intermediate (21). Films of RuPSSA made by the ion exchange of Ru(III) with HPSSA achieved approximately 75% exchange.

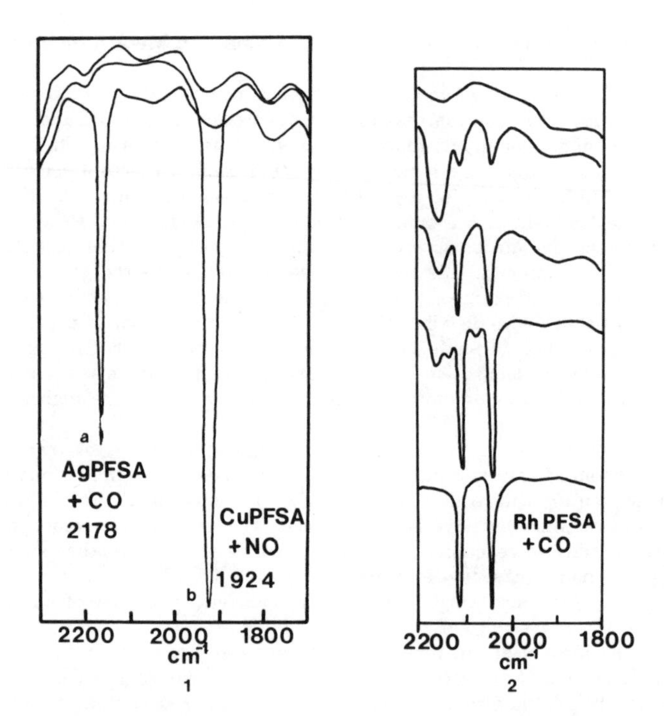

Figure 1. The infrared spectra of AgPFSA (a), AgPFSA exposed to CO (b), and Cu(II)PFSA exposed to NO.

Figure 2. Infrared spectra of RhPFSA exposed to CO at increasing pressure, temperature and evacuation.

These films were reacted in several sequences with CO and H_2, as well as ^{13}CO and D_2 to permit identification of the species formed by their infrared spectra. This is of particular value since both ν(CO) and ν(Ru–H) are expected in the 1900–2200 cm^{-1} region. As summarized in Fig 3, treatment of RuPSSA with these gases lead to identification of both carbonyl and hydride species. The identification of the particular carbonyls should be helpful in studying Ru-based catalysts, whose ν(CO) spectra are complex. The identification of the hydride species, confirmed by deuteration, is novel. The figure also shows the positions of ν(CO) bands resulting from the reaction of CH_3OH with RuPSSA.

Many transition metal ionomers can be reduced to the metallic form. This has been accomplished in several ways, including reductive deammination in the case of $Pt(NH_3)_4^{+2}$ PSSA (16), reduction with hydrazine followed by reductive deammination (15), reaction with H_2 (15–18,20), and effectively reduction with H_2 formed by the water gas shift reaction of CO with included H_2O (14–18). The products of these reductions, as well as these products after subsequent reaction with O_2 or CO, have been studied by electron microscopy. While space does not permit a complete presentation of the results, two important conclusions can be noted. In the cases of Ag, Pt, Rh and Ru in PFSA ionomers, the metal particles are relatively small and have a quite sharp particle size distribution (17,20,22,23). Since sample handling methods require their exposure to air after formation, the particles should be considered to be oxidized to some extent. The particle size distributions for these particles are shown in Fig 4. All of the distributions are peaked in the 20–40 A° diameter range. Since the metals, the degree of initial exchange, and the method of treatment are all quite different, we take this to mean that the PFSA morphology was not changed significantly by the mild solution exchange process and that it exerted control over the size of the particles formed upon reduction. That control was not complete, since the materials were heated during treatment and some distribution of domain sizes is expected in any event, but it must have been extraordinarily effective to result in particles with both a narrow size distribution and an average value that is close to that expected for the size of PFSA domains.

The other conclusion reached from such studies is that only very small metal particles are formed under our conditions when RhPSSA or RuPSSA are reduced (19,21). Within our detection limits, the metal is uniformly distributed in metal particles of significantly less than 10 A° size. Since the HPSSA ionomer that also results from the reduction is soluble, this may provide a method for obtaining a dispersion of small Rh or Ru particles.

Finally, several of the metal ionomer systems have been tested for their catalytic activity (17,22,23). These studies, also, are too extensive to detail here, but several results can be stated. Of particular importance, the reduced RuPFSA, RhPFSA and PtPFSA materials catalyze the oxidation of CO effectively. The order of CO oxidation rates is RuPFSA > RhPFSA > PtPFSA. The rate of the $2CO/O_2$ reaction is significantly lower with PtPFSA than it is with RuPFSA and RhPFSA, but its activation energy with PtPFSA is similar to that calculated for known SiO_2–Pt catalysts. The fact that the rates and apparent activation energies for the reactions carried out with reduced RhPFSA and RuPFSA are lower than those over SiO_2–supported catalysts shows that the reactions are gas diffusion limited.

Particle Formation and Reaction Kinetics: Silver(I) PFSA Reduction and Oxidation

A system of controlled size, supported silver particles, such as Ag–PFSA, could be convenient for the study of chemisorption reactions, particularly those of O_2, because of their importance in catalytic oxidations. Experiments involving the formation of silver particles upon reduction of Ag^+–PFSA and their subsequent exposure to oxygen, will be discussed here.(20)

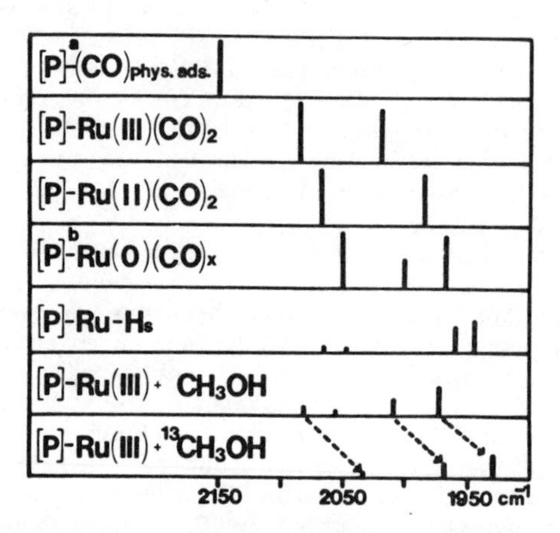

Figure 3. Summary representation of infrared bands observed from carbonyl and hydride species formed on RuPSSA by treatments with CO, H_2, O_2 and CH_3OH. Reproduced with permission from Ref. 21, Copyright 1985, Academic Press, Inc.

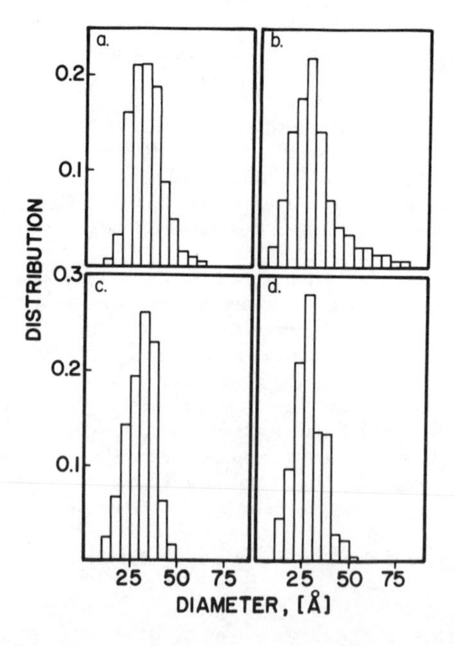

Figure 4. Particle size distributions in (a) PtPFSA, (b) AgPFSA, (c) RuPFSA, and (d) RhPFSA.

Exposure of Ag^+-PFSA to H_2 (1 atm, 300 K) was monitored by IR spectroscopy. The spectral area between 1600 and 1900 cm^{-1} is of interest since the water associated with protonated sulfonic groups can be differentiated from the water coordinated to Ag^+-sulfonic groups by their bending modes occuring in this region (1710 and 1620 cm^{-1}, respectively, Fig 5a). Exposure of Ag^+-PFSA to H_2 yields an increase in the 1710 cm^{-1} (see Fig 5b) band and this shows that reduction of Ag^+ takes place under those conditions. During the reduction procedure the initially colorless membranes attain a yellow color. As shown in Fig 6, noticeable changes in the visible absorption spectrum occur after several hours of exposure to H_2, when the film is characterized by an asymmetric absorption envelope in the 300–400 nm region. Thus we correlate their absorption features in the visible with the formation of silver particles upon reduction of Ag^+-PFSA. A number of investigations of small silver particles (10–100 Å in diameter) supported on a variety of substates yielded similar absorption characteristics in the visible. They are thought to be due to collective electronic excitations (plasmon modes).

Exposure of the membranes prepared this way to oxygen by admitting air into the cell, inverts the infrared observations (Fig 5c), a clear sign that metallic silver is *partially* oxidized and becomes associated with the sulfonic group. The changes in the visible spectrum are much more dramatic. A very intense band grows at *ca* 370 nm and the membranes attain a bright golden color. The evolution of the new band is shown in Fig 7, for a film that has been reduced for 19 hrs. (a) and then exposed to air for 10 (b), 20 (c), 30 (d), 40 (e) and 500 (min). This optical absorption is assigned to transitions that transfer charge from bulk Ag^0 to Ag^+ ions that are adjacent to electronegative species such as oxygen atoms. Theoretically such charge transfer transitions for species of the type (Ag–Ag^+–0) were predicted to occur in the 350–380 nm range. If the above assignment is correct, the visible spectra of Ag^0-PFSA exposed to air could reveal the details of oxygen chemisorption on the PFSA supported Ag particles. In Fig 8, the evolution of Absorbance at 370 nm (A_{370}) versus time of exposure to air is presented for Ag-PFSA membranes prepared after various times of reduction of Ag^+-PFSA. Such kinetic data could be modelled by a function for the reaction rate such as $dA_t/dt = A_\infty K\exp(-kt)$ where A_t is the optical absorbance at time t, A_∞ that at the completion of the reaction, K a constant and k the rate constant.

Thus we tested the following rate law:

$$\ln\left(1 - \frac{A_t}{A_\infty}\right) = -kt \qquad (1)$$

A plot of Eq 1 for the case of the membrane prepared by 19h of reduction is shown in Fig 9, for the graphical determination of k. It can be seen that although our model describes very well our data for times greater than 40 min., in the short time scale a second faster process is involved. This is a common result of all the films we studied. We found that in all cases our absorption versus time data fitted a curve of the form:

$$A_t = A_{\infty,1}(1 - \exp(-k_1t)) + A_{\infty,2}(1 - \exp(-k_2t)) \qquad (2)$$

A representative fitting corresponding to Eq 2 is shown in Fig 10. The k_1 and k_2 values we calculated from Eq 2 by least squares fitting with a corelation coefficient higher than 0.999 are listed in Table 1. All k_1 values are of the order of 10^{-4} s^{-1} where all k_2 ones are about 10^{-3} sec^{-1}. The average value of the ratios k_2/k_1 is 10.0.

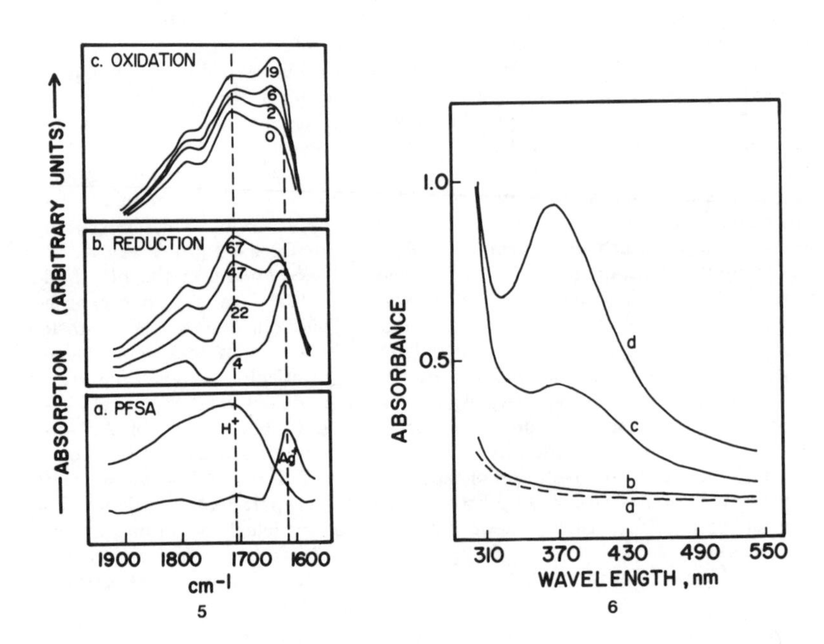

Figure 5. Infrared spectra of HPFSA and AgPFSA subjected to the following treatments: (a) HPFSA and AgPFSA; (b) AgPFSA films reduced with H_2 at 1 atm and 300K for 4, 22, 47 and 67 hours; (c) AgPFSA films that had been reduced for 67 hours in H_2 and subsequently exposed to air for 0, 2, 6, and 19 hours. Reproduced with permission from Ref. 20, Copyright 1985, Academic Press, Inc.

Figure 6. Visible absorption spectra of AgPFSA exposed to H_2 for (a)0 time, (b) 14 hours, (c) 19 hours and (d) 24 hours. Reproduced with permission from Ref. 20, Copyright 1985, Academic Press, Inc.

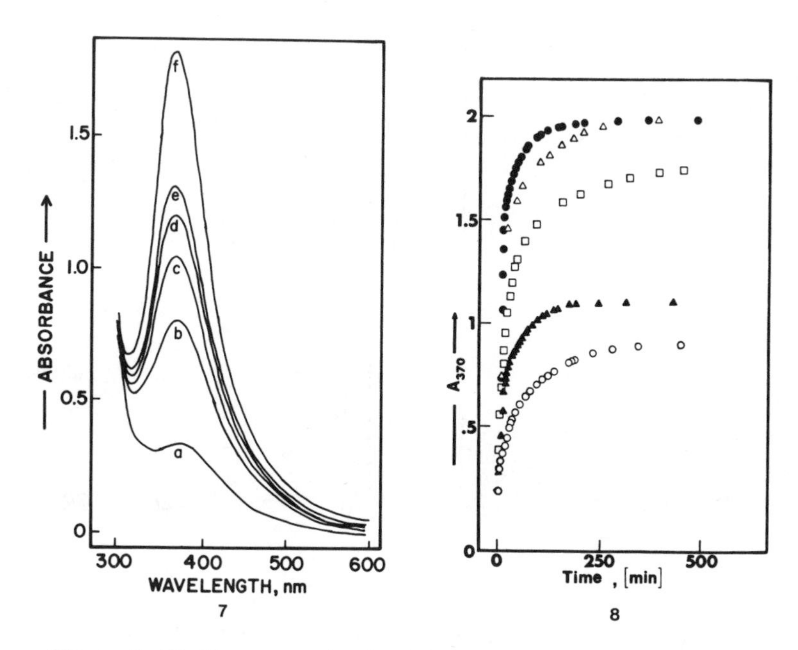

Figure 7. Visible spectra of reduced AgPFSA exposed to air for (a) 0, (b)10 min., (c) 20min., (d) 30 min., (e) 40 min. and (f) 500 min. Reproduced with permission from Ref. 20, Copyright 1985, Academic Press, Inc.

Figure 8. The time evolution of absorbance at 370 nm exhibited by reduced AgPFSA films exposed to air. The different curves correspond to membranes prepared after 9 (filled triangles), 14 (open circles), 19 (open squares), 23 (open triangles) and 24 (filled circles) hours of exposure to H_2. Reproduced with permission from Ref. 20, Copyright 1985, Academic Press, Inc.

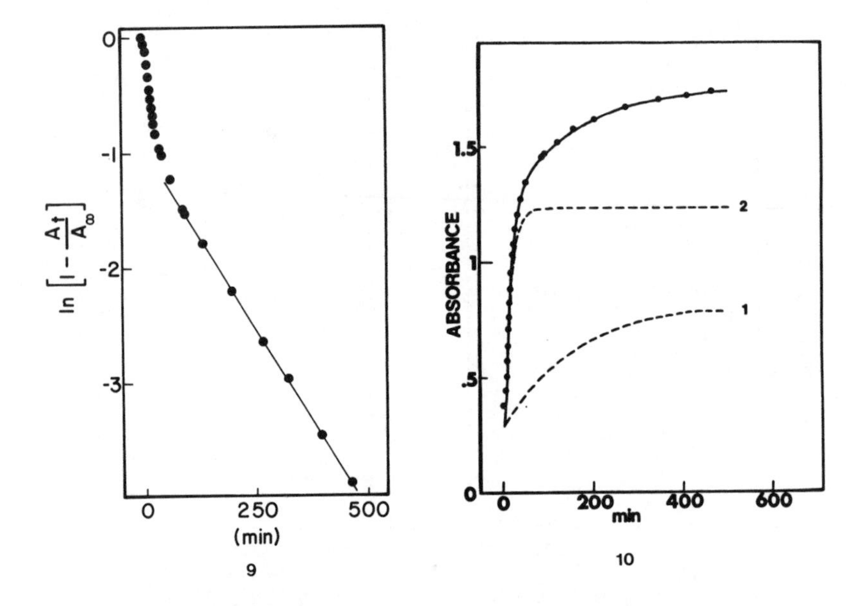

Figure 9. The $\ln[1 - (A_t/A_\infty)]$ versus time for a reduced AgPFSA film exposed to air after 19 hours of reduction in H_2. Dots represent points calculated from experimental data. The solid line shows the least squares fitting to a straight line on the long time scale. Reproduced with permission from Ref. 20, Copyright 1985, Academic Press, Inc.

Figure 10. The deconvolution of the absorbance at 370 nm, A_{370}, versus time of exposure to air plot in the case of a reduced AgPFSA film prepared by 19 hours of reduction. Dots represent the experimental data shown in Fig. 8. Dashed lines are the deconvoluted slow (1) and fast (2) reaction curves. The sum of these is the solid line. Reproduced with permission from Ref 20, Copyright 1985, Academic Press, Inc.

Table 1. Values of the Parameters of Eq. 2 for the Ag(0)PFSA Films

Time of reduction (hours)	$k_1 (s^{-1})$	$k_2 (s^{-1})$	$\frac{k_2}{k_1}$
9	2.8×10^{-4}	3.47×10^{-3}	12.39
14	1.1×10^{-4}	0.85×10^{-3}	7.73
19	1.1×10^{-4}	1.08×10^{-3}	9.82
23	2.0×10^{-4}	2.02×10^{-3}	10.10
24	3.5×10^{-4}	3.42×10^{-3}	9.77
		Average:	9.96

A detailed discussion of the two parallel reactions is beyond the scope of this paper on ionomers, but it is appropriate to note that the fast and slow reactions are tentatively assigned to chemisorptions of atomic oxygen and molecular oxygen (O_2^-), respectively, on the surface of silver particles in PFSA.

Overall, this study and the others mentioned in Sections C and D illustrate that ionomers provide supports for a variety of interesting chemical reactions, including metal particle formation, which can be either quite different from those in other media or they can be sufficiently similar to carry out known reactions in a useful type of polymeric material.

Acknowledgments

The partial support of this work and the use of the facilities of the Materials Research Laboratory and the support of the Office of Naval Research are gratefully acknowledged.

Literature Cited

1. A. Eisenberg and M. King, *Ion-Containing Polymers*, Volume 2, Polymer Physics. Academic Press, New York, San Francisco, London, 1977.
2. A. Eisenberg, *Ions in Polymers*, Americal Chemical Society, Washington, D.C., 1980.
3. T. D. Gierke, G. E. Munn and F. C. Wilson, *J. Polym. Sci. Phys.* 1981, **19**, 1687.
4. Ogumi, Z., Takehara, Z., Yoshizawa, S., *J. Electrochem. Soc. - Electrochem. Sci. and Technol.* 1984, **131**, 769.
5. L. R. Falkner, *Chem. and Eng. N.*, February 27, 1984, and references therein.
6. N. E. Prieto and C. R. Martin, *J. Electrochem. Soc. - Electrochemical Sci. and Technol.* 1984, **131**, 751.
7. J. Rubinstein and A. J. Bard, *J. Amer. Chem. Soc.*, 1980, **102**, 6641.
8. D. A. Buttry and F. C. Anson, *J. Amer. Chem. Soc.* 1984, **106**, 59.
9. G. A. Olah, J. Kaspi, and J. Bukala, *J. Org. Chem.* 1977, **42**, 4187.
10. F. J. Waller, *Catalytic Conversions of Synthesis Gas and Alcohols to Chemicals.* (R. G. Herman, Editor), Plenum Publ. Corp., 1984.
11. F. J. Waller, U. S. Patent 4, 414, 409, (1983).
12. W. H. Kao and T. Kuwana, *J. Amer. Chem. Soc.*, 1984, **106**, 473.
13. Bailar, J. G., Jr., *Cat-Rev-Sci. Eng.* 1984, **10**, 17.
14. S. L. Peluso, Ph.D. Thesis, Brown University, 1980.
15. D. M. Barnes, Ph.D. Thesis, Brown University, 1981.
16. S. Noor Chaudhuri, Ph.D. Thesis, Brown University, 1982.
17. V. D. Mattera, Jr., Ph.D. Thesis, Brown University, 1984.
18. I. W. Shim, Ph.D. Thesis, Brown University, 1985.
19. V. D. Mattera, Jr., P. J. Squattrito and W. M. Risen, Jr., *Inorg. Chem.* 1984, **23**, 3597.
20. G. Chryssikos, V. D. Mattera, Jr., A. T. Tsatsas, and W. M. Risen, Jr., *J. Catal.* (1985) **93**, 430.
21. I. W. Shim, V. D. Mattera, Jr., and W. M. Risen, Jr., *J. Catal.* (1985) **94**, 531.
22. V. D. Mattera, Jr., S. N. Chaudhuri, R. Gonzalez, and W. M. Risen, Jr., in preparation.
23. V. D. Mattera, Jr., D. M. Barnes, R. Gonzalez, and W. M. Risen, Jr., submitted *J. Catal.* (1984).
24. J. T. Yates, *J. Chem. Phys.* 1979, **71**, 3908.

RECEIVED December 13, 1985

6

Studies on the Synthesis of Novel Block Ionomers

Robert D. Allen, Iskender Yilgor, and James E. McGrath

Department of Chemistry and Polymer Materials and Interfaces Laboratory, Virginia Polytechnic Institute and State University, Blacksburg, VA 24061

Novel sulfonated and carboxylated ionomers having "blocky" structures were synthesized via two completely different methods. Sulfonated ionomers were prepared by a fairly complex emulsion copolymerization of n-butyl acrylate and sulfonated styrene (Na or K salt) using a water soluble initiator system. Carboxylated ionomers were obtained by the hydrolysis of styrene-isobutylmethacrylate block copolymers which have been produced by carefully controlled living anionic polymerization. Characterization of these materials showed the formation of novel ionomeric structures with dramatic improvements in the modulus-temperature behavior and also, in some cases, the stress-strain properties. However no change was observed in the glass transition temperature (DSC) of the ionomers when compared with their non-ionic counterparts, which is a strong indication of the formation of blocky structures.

Ion containing polymers (ionomers) are receiving ever increasing attention due to the dramatic effects that small amounts of ionic groups exert on polymer properties (1-4). Increases in tensile strength, modulus, melt and solution viscosity, glass transition temperature, and a broadening of the rubbery plateau are some of the major changes that occur when ionic groups are incorporated into polymers. The extent of property modification depends generally on the type, amount, and distribution of ionic groups in the polymer. Although utilization of a wide variety of ionic groups is possible, carboxylates and sulfonates have thus far received considerable attention. Polymer backbones modified most widely include polyethylene, polystyrene, poly(acrylates), polypentenamers, ethylene-propylene-diene terpolymers and polysulfones (4-9). However, the major difficulty in this field is the lack of available synthetic methods that would lead to the preparation of well-defined ionomeric structures.

0097-6156/86/0302-0079$06.00/0

Ion containing polymers are prepared via two fundamentally different approaches. The first method involves a post-polymerization reaction in which the ionic group is "grafted" onto a preformed polymer (4,10,11), whereas in the second method a vinyl monomer is directly copolymerized with an unsaturated acid or salt derivative. Another possibility, which in fact is a combination of the above two approaches, is to copolymerize two covalent monomers and then derivatize one by, for example, ester hydrolysis (12). Carboxylated ionomers are generally synthesized by the direct copolymerization of acrylic or methacrylic acid with vinyl monomers. The copolymers produced are then either partially or completely neutralized with bases to incorporate the various counterions (e.g. Na^+, K^+, Zn^{++}, etc.) into the system. Sulfonate groups are most often introduced into the polymer backbone via post-polymerization reactions using sulfonating agents such as acetyl sulfate or SO_3/triethylphosphate complex in a suitable solvent (4,10,11). The sulfonic acid groups introduced into the polymer backbone are then neutralized. This method of sulfonation is difficult and is limited to only backbones which provide suitable sites for sulfonation. Therefore, although sulfonated ionomers reportedly exhibit association properties far superior to carboxylated systems (4), relatively limited work appears in the literature on the direct synthesis of sulfonated ionomers from the respective monomers (13,14).

The synthetic routes discussed above usually result in random placement of ionic groups along the polymer backbone, thus making characterization and interpretation of structure - property studies quite difficult. Previous work on the synthesis of well-defined block and graft ionomers based on polystyrene and poly(2-or 4-vinylpyridine) have been done by Selb and Gallot (15,16). They have synthesized these copolymers via anionic polymerization at -70°C using diphenylmethylsodium as the initiator in THF. They have also investigated the solution behavior of the resulting ionomers after quaternization (17). More recently Gauthier and Eisenberg (30) have demonstrated the synthesis of styrene-vinylpyridinium ABA block ionomers and studied their thermal and mechanical properties. Recent work with so-called "telechelic" ionomers, where the ionic groups are located only at the chain ends, may also simplify this characterization problem (18-21).

We are currently exploring new routes to the synthesis of ionomers with controlled architecture, i.e. with control over the amount and location of ionic groups in the polymer backbone. One of our main interests is the synthesis of ion containing block copolymers. The applicability of anionic polymerization in the synthesis of block copolymers and other well defined model systems is well documented (22-24). Not as well appreciated, however, is the blocky nature that certain emulsion copolymerizations may provide. Thus, we have utilized both anionic and free radical emulsion polymerization in the preparation of model ionomers of controlled architecture. In this paper, the synthesis and characteristics of sulfonated and carboxylated block ionomers by both free radical emulsion and anionic polymerization followed by hydrolysis will be discussed.

EXPERIMENTAL

A) Emulsion Copolymerization

Sodium-p-styrene sulfonate (SST) was supplied by Exxon Research and Engineering Co. and used without further purification. n-Butyl acrylate was a product of Polysciences, Inc. and was passed through a column of neutral alumina, twice, before use to remove the inhibitor. Potassium persulfate and sodium bisulfate were recrystallized from distilled water and dried under vacuum. Nonionic emulsifier (ATLOX 8916 TF) was a product of ICI Americas Inc. 1-Dodecanethiol (Aldrich) was used as the chain transfer agent. Double distilled and deoxygenated water was used throughout the emulsion reactions.

Copolymerizations were carried out in a 4-necked, 250 mL round bottom flask fitted with a mechanical stirrer, condenser, thermometer, and gas (N_2) inlet. Temperature was controlled in the 60-70°C range and the reaction times were varied between 5 and 24 hours. The emulsion recipe used is given in Table I.

TABLE I

Typical Emulsion Copolymerization Recipe

Water	80 mL
Vinyl Monomer	20 g
Sulfonated styrene	0-2 g
Emulsifier	0.8 g
$K_2S_2O_8/NaHSO_3$	0.08/0.04g
1-Dodecanethiol	0-0.1 g

At the end of the reactions the stable latex obtained was poured into teflon coated aluminum pans and dried in an air oven. The products were then dissolved in THF and coagulated in a methanol/water mixture several times, before final drying.

B) Anionic Copolymerization

Anionic techniques were used to synthesize the block polymers which function as precursors to ion containing block polymers. Monomers were carefully dried by repeated vacuum distillation from CaH_2. Distillation from dibutyl magnesium was also utilized for the final purification of the hydrocarbon monomers, styrene and diphenyl ethylene. The methacrylate monomers may also be finally purified via the trialkyl aluminum approach (25), although this was not necessary as the Rohm & Haas iso-butyl methacrylate (IBMA) used in this study proved to be exceedingly pure. THF was dried by double distillation from sodium/benzophenone complex.

Block polymerizations were conducted in THF at -78°C under an inert atmosphere. The polymerization route employed is shown in Scheme I. Typically, the styrene monomer was charged into the polymerization reactor with THF, followed by rapid addition of sec-butyl lithium. The polystyryl lithium was then "capped" with 1,1-diphenyl ethylene to form less basic, more hindered anions so as to avoid deleterious side reactions to the methacrylate carbonyl. Just before the addition of methacrylate monomer (IBMA), a small amount of capped polystyryl lithium was removed for GPC analysis. Termination was performed several minutes after the addition of the methacrylate monomer, using a methanol/acetic acid mixture. The polymers were generally isolated by precipitation in methanol. During the synthesis reactions molecular weights ($\bar{M}n$) of the polystyrene and poly(iso-butyl) methacrylate) blocks were kept constant at 40,000 and 10,000 g/mole respectively.

After drying this block polymer precursor, partial hydrolysis of the methacrylate block lead to the ion containing block copolymer. The hydrolysis route is shown in Scheme II. The hydrolysis method employed utilized potassium superoxide as a general route to ester cleavage to generate in a direct fashion potassium carboxylate units.

Typically, the copolymer is dissolved in a mixed solvent system such as THF/DMSO or Toluene/DMSO. A three fold excess of powdered KO_2 (Alfa) is then added. Reaction time and temperature depend on degree of cleavage desired as well as the type and nature of the ester alkyl group. The reaction is quenched with water. The ion containing block copolymer is then reprecipitated and dried.

Characterization of Products

Intrinsic viscosities were determined in THF at 25°C using Ubbelohde viscometers. Molecular weights were also analyzed using a Waters 150C GPC with microstyragel columns, although Bondagel columns were used in some experiments. Thermal characterization of the products were performed on a Perkin Elmer Thermal Analysis System 2. Tg's were determined by DSC with a heating rate of 10°C/min. TMA penetration curves were obtained with a constant load of 10g and heating rate of 10°C/min. Stress-strain tests were carried out using an Instron Model 2211 Tester. Dog-bone shaped samples were punched out of compression molded polymer films using standard dies. Tests were performed at room temperature with a crosshead speed of 10 mm/min. Infrared spectra were obtained on a Nicolet MX-1 FT-IR Spectrometer.

RESULTS AND DISCUSSION

A) Emulsion Copolymerization of n-Butyl Acrylate and Sulfonated Styrene

We have investigated the direct copolymerization of sulfonated

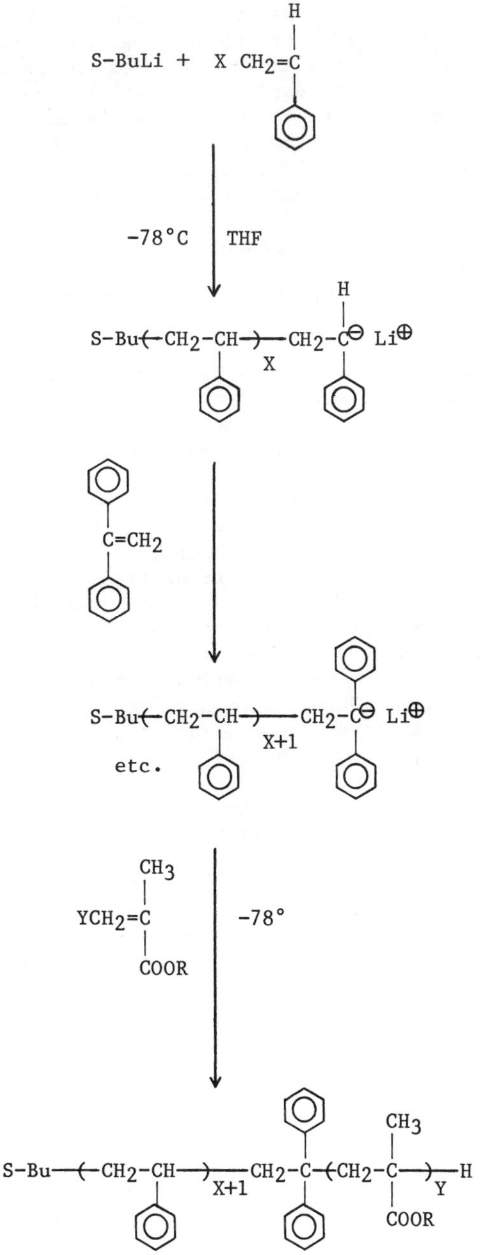

$$\text{S-BuLi} + X\ CH_2{=}C(H)(C_6H_5)$$

$$\xrightarrow[\text{THF}]{-78^\circ C}$$

$$\text{S-Bu}\!\left(\!-CH_2-CH(C_6H_5)-\!\right)_X\!-CH_2-C^{\ominus}(H)(C_6H_5)\ Li^{\oplus}$$

$$\xrightarrow{(C_6H_5)_2C{=}CH_2}$$

$$\text{S-Bu}\!\left(\!-CH_2-CH(C_6H_5)-\!\right)_{X+1}\!-CH_2-C^{\ominus}(C_6H_5)_2\ Li^{\oplus}$$

etc.

$$\xrightarrow[-78^\circ]{Y\,CH_2{=}C(CH_3)(COOR)}$$

$$\text{S-Bu}\!-\!\left(\!-CH_2-CH(C_6H_5)-\!\right)_{X+1}\!-CH_2-C(C_6H_5)_2\!-\!\left(\!CH_2-C(CH_3)(COOR)-\!\right)_Y\!-H$$

SCHEME I

Anionic Synthesis of Styrene/IBMA Block Copolymer

styrene (SST) with n-butyl acrylate to produce ionomers. It was demonstrated that emulsion polymerization is a useful and potentially important technique in the synthesis of sulfonate ion containing polymers by direct reaction between the ionogenic and covalent monomers, provided that a "water soluble initiator system" is used. Some workers have already reported the use of an organic soluble peroxide and a water soluble reducing agent during similar reactions in order to generate effective copolymerization at the micelle interfaces (13,14).

During our studies we used both sodium and potassium salts of SST and they behaved similarly as expected. The emulsifier was a nonionic "alkylphenoxy-ethylene oxide" type surfactant with an HLB value of 15.4. The amount of SST charged was varied between 0 and 10 percent by mole. The latex obtained after the reaction was fairly stable in each case. Table II provides the data on the synthesis of n-butyl acrylate and sulfonated styrene ionomers. As expected in emulsion polymerizations we were able to obtain fairly high yields. In the first four reactions no chain transfer agent (CTA) was used and except sample 1, which is the control poly(n-butyl acrylate), we were not able to dissolve any others (samples 2,3 or 4) in any single solvent or solvent combinations. This is attributed to the combined effects of very high molecular weights of the polymers produced and strong ionic interactions between the pendant sulfonate groups on the polymer chains and possibly the residual end groups derived from the initiator. Although these materials were not soluble, they were easily compression moldable at temperatures around 200°C and all yielded clear, transparent films. However, when a small amount of an organic soluble chain transfer agent (1-dodecanethiol) was used, it was possible to obtain ionomers which were soluble in solvents such as tetrahydrofuran, chloroform and toluene/methanol.

As can be seen from the last column of Table II, DSC studies did not indicate any change in the glass transition temperature of the polyacrylates due to the presence and/or concentration of the ionogenic monomer (SST) incorporated. We have observed the same behavior in styrene, methyl acrylate and other systems (26). Other workers have also reported similar results in ionomers synthesized by the direct reactions of various acrylic acid salts and covalent vinyl monomers (27). This indicates either the simultaneous homopolymerization of both monomers or a "block" copolymerization. We were however unable to detect any high temperature transitions in DSC due to the very small amounts of SST incorporated and the sensitivity of DSC method. In methyl acrylate/SST system at 8 mole percent SST level we were able to detect a second Tg at 330°C for the ionic block (28). However when we studied the thermomechanical behavior of these materials via TMA in the penetration mode, we were able to see the dramatic effects of ion incorporation even at very low levels, on the highly extended "rubbery plateau" in SST containing polymers compared to the control poly(n-butyl acrylate). This is illustrated in Figure 1. Attempts to physically blend the two homopolymers showed that they were highly incompatible as expected and there was no enhancement in the thermal or mechanical properties of resulting materials. The films obtained by blending were very cloudy, showing the dispersion of poly(sulfonated styrene)

SCHEME II

Synthesis of Ion-Containing Block Copolymers via Superoxide Hydrolysis

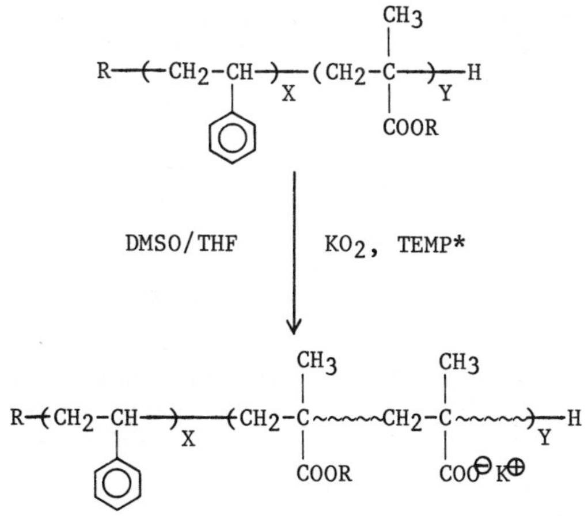

$$R\text{-}(CH_2\text{-}CH(C_6H_5))_X\text{-}(CH_2\text{-}C(CH_3)(COOR))_Y\text{-}H$$

DMSO/THF KO_2, TEMP*

$$R\text{-}(CH_2\text{-}CH(C_6H_5))_X\text{-}(CH_2\text{-}C(CH_3)(COOR)\sim\sim CH_2\text{-}C(CH_3)(COO^{\ominus}K^{\oplus})\sim\sim)_Y\text{-}H$$

*R=Me, T=25°C
R=i-Butyl, T=85°C

TABLE II

Synthesis of n-Butyl Acrylate Sulfonated Styrene Copolymers

No	nBuA (g)	SST (g)	SST (%mol)	Reaction T,°C	Reaction t,hr	CTA (g)	Yield (%)	Tg °C
1	10.0	--	--	60	19	--	85	-55
2	10.0	0.52	2.8	65	9	--	95	-55
3	10.0	1.01	5.5	65	9	--	93	-55
4	10.0	1.90	9.8	60	5.5	--	93	-55
5	11.0	0.99	5.5	67	24	0.13	82	-55
6	25.0	0.40	1.0	62	5	0.09	91	-55
7	24.2	0.80	2.0	68	5	0.09	72	--
8	23.3	1.12	2.9	68	5.5	0.05	65	--
9	23.3	1.50	3.8	68	5.5	0.05	68	-55

2-5, K^+ salt, # 6-9, Na^+ salt.

in polyacrylate. On the other hand compression molded films of the copolymers synthesized appeared macroscopically "homogeneous" and transparent, all of which indicates the formation of "blocky" structures. Our kinetic studies on the copolymerization of SST and methyl acrylate in emulsion also show the formation of "blocky" ionomers (28).

Stress-strain behavior of n-butyl acrylate-SST ionomers are given in Figure 2. It is clear that the incorporation of sulfonate groups has a remarkable effect on the behavior of resulting materials. Compared with n-butyl acrylate homopolymer, introduction of approximately 5 mole percent of SST into the copolymer increases the tensile strength more than 40-fold. The increase in the overall strength is proportional to the level of ion content. Modulus values of the resulting ionomers increase and elongation at break slightly decreases with increasing ion content. TMA and stress-strain results may indicate that in this system optimum properties can be obtained at a SST level of about 5 mole percent.

B) Block Ionomers Through Anionic Polymerization and Hydrolysis

The novel reaction scheme which we had proposed for "blocky" ionomers through emulsion polymerization prompted us to prepare very well defined ion containing block copolymers through anionic polymerization. The initial system chosen for synthesis, modification, and characterization was the poly-styrene-poly(iso-butyl methacrylate) diblock (PS-PIBM DB) copolymer. This particular system was chosen for the following reasons:

1) The availability, ease of purification, and well established polymerization characteristics of styrene.

2) The large body of literature available on polystyrene based ionomers.

3) IBMA (Rohm and Haas) is a methacrylate monomer of high purity.

4) The methacrylate ester group may be converted, via hydrolysis, to a carboxylate ion.

These polystyrene-methacrylate copolymers with block molecular weights of 40,000 and 10,000 g/mole respectively, are thus precursors to carboxylate ion containing block copolymers.

As is evident from the GPC chromatograms in Figure 3, both the diphenyl ethylene capped polystyrene and the PS-PIBM diblock copolymers have narrow molecular weight distributions. More importantly, no detectable homopolymer contamination is present in the very pure diblock. This high structural integrity was achieved by taking the following precautions.

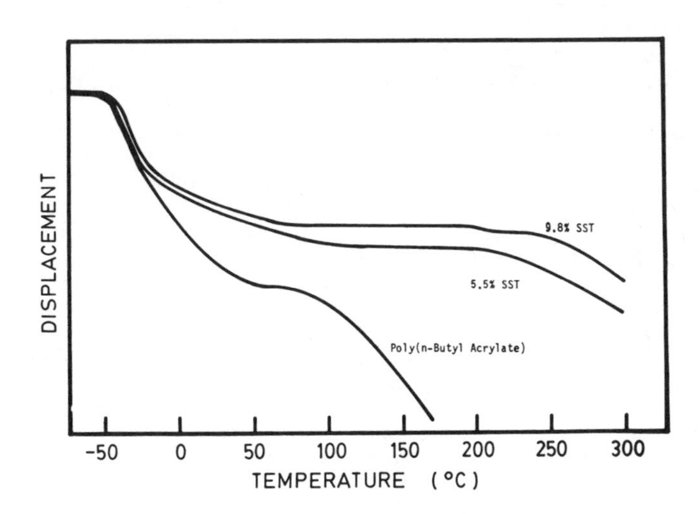

Figure 1. TMA Penetration Curves for n-Butyl Acrylate/Sulfonated Styrene Ion-Containing Copolymers.

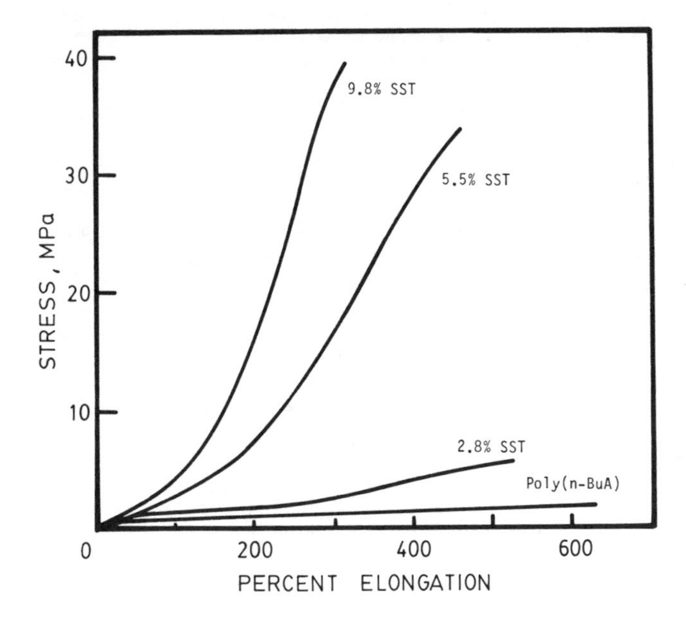

Figure 2. Stress-strain Behavior of n-Butyl Acrylate/Sulfonated Styrene Ion-Containing Copolymers.

1) Capping of the PS "Front Block" with 1,1-DPE to afford a highly hindered, less basic macromolecular initiator for methacrylate polymerization.

2) Slow addition at -78°C of IBMA to avoid thermal termination.

3) Utilization of high purity methacrylate monomers.

Hydrolysis of this well defined diblock material proved to be quite difficult. A variety of acidic and basic hydrolysis attempts were made, with little or no success. We then found evidence in the literature (29) for a facile new hydrolysis method which had achieved very fast hydrolysis of methyl laurate by using solid potassium superoxide (KO_2). We have utilized this new method successfully in the hydrolysis of polymethyl methacrylate (PMMA). Elvacite 2041 (DuPont PMMA) was rendered water soluble by simply stirring over KO_2 at room temperature overnight. Scheme II shows the hydrolysis route employed. As is indicated in the scheme, although PMMA is hydrolyzed at room temperature, the bulkier iso-butyl ester group is much more resistant to this hydrolysis reaction, and indeed, even at 80°C the hydrolysis is slow as judged by FT-IR. Figure 4 shows a typical FT-IR spectrum of the partially hydrolyzed diblock in the acid form. The extent of hydrolysis vs. reaction time is shown in Table III where the absorbance ratio of acid to ester is given. As can be seen, there appears to be an induction period in this hydrolysis reaction. This may be due to the heterogeneous nature of KO_2 in THF/DMSO. We have also carried out these reactions in toluene/DMSO solvent mixture which quite surprisingly gives a much more homogeneous reaction mixture. These KO_2 reactions performed in toluene/DMSO solvent systems show no induction period, thus leading to much shorter reaction times. It should also be noted that no degradation of the polymer resulted from these reaction conditions as noted from GPC analysis on the acidified "ionomers", using bondagel columns and THF solvent at room temperature.

Characterization of the thermal and mechanical properties of these systems show several marked similarities with the poly(butyl acrylate sulfonated styrene) emulsion polymers discussed earlier. Transparent films were produced by compression molding under high pressures at 250°C. These ion containing block copolymers were insoluble in all solvents tried, but dissolve quickly when several drops of HCl are added to convert the carboxylate ions to carboxylic acid groups. As seen in Figure 5, the block ionomer having 10 mole percent ionic content shows a very highly extended rubbery plateau in the TMA experiment compared to the diblock precursor. At the same time, dynamic mechanical analysis (DMTA) shows no change in the glass transition behavior of the polystyrene matrix with ion content, as well as highly extended rubbery plateau behavior with a very high modulus. These two results are remarkably similar to the very differently synthesized butylacrylate-sulfonated styrene ionomers via emulsion copolymerization.

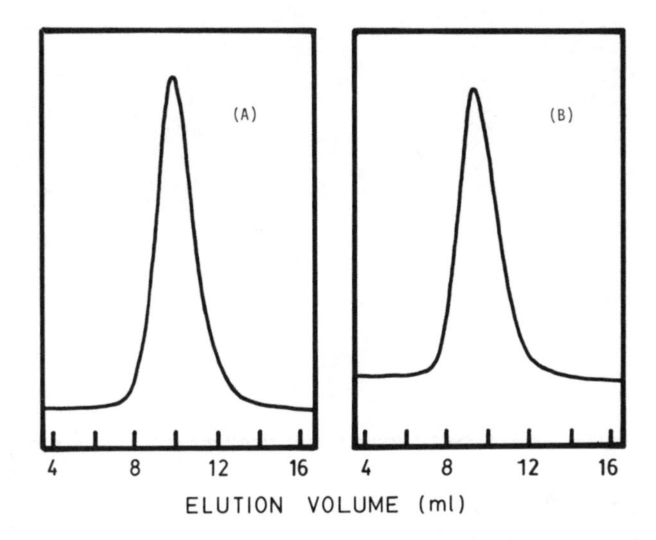

Figure 3. Size exclusion chromatograms of (A) polystyrene front block ($\overline{M}n$ 40,000 g/mole) and (B) polystyrene/poly(isobutyl methacrylate) diblock copolymer ($\overline{M}n$ 50,000 g/mole).

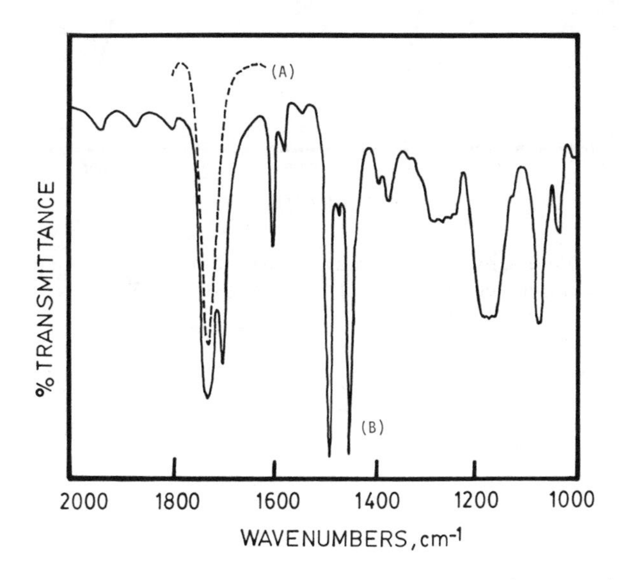

Figure 4. FT-IR spectra of (A) polystyrene/poly(isobutyl methacrylate) diblock copolymer and (B) carboxylated block ionomer obtained after hydrolysis with KO_2.

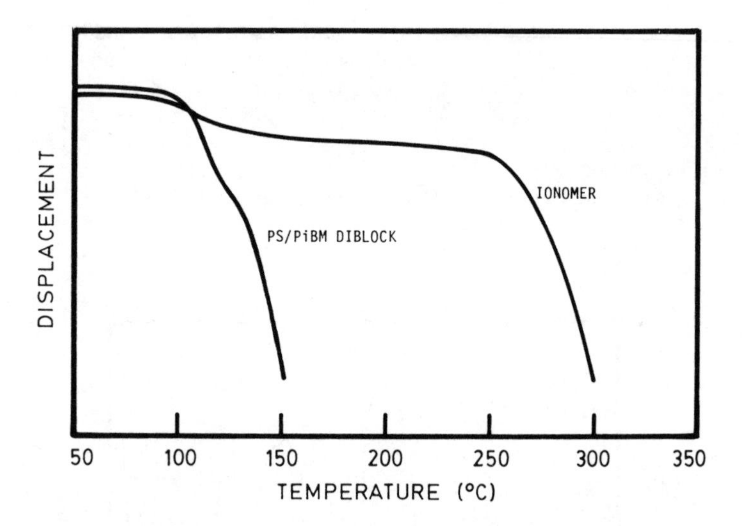

Figure 5. TMA Penetration Behavior of PS/PiBM Diblock Copolymer and Carboxylated Ionomer (5 mole percent-$COO^{\ominus}$ $K^{\oplus}$) Obtained after hydrolysis.

TABLE III

Studies on Superoxide Hydrolysis and Characterization of the Products

Sample No.	Rxn. Time (Hr.)	El. Vol. (GPC)	A_A/A_E (FT-IR)
I	0	28.8	0
II	1	28.7	0
III	2	28.8	0
IV	6	28.8	0.44
V	13	28.8	0.64
VI	17	29.3	1.15

CONCLUSIONS

Both emulsion copolymerization of butyl acrylate/sulfonated styrene with a water soluble initiation system, and anionically synthesized diblock ionomers based on the ester hydrolysis of methacrylate blocks show novel and quite similar characteristics. Both systems show no effect of ion content on the Tg of the matrix however provide a highly extended rubbery plateau to the resulting ionomers.

Our work on the synthesis of model ionomer systems by anionic polymerization is continuing. The architectural effects of these ionomers having controlled structures are being compared with their random counterparts.

ACKNOWLEDGMENT

The authors would like to thank the Petroleum Research Fund (Grant No. 14747-AC7) and the Exxon Foundation for sponsoring this research.

Literature Cited

1. L. Holliday, Ed., "Ionic Polymers", Applied Science Publishers, London 1975.
2. A. Eisenberg and M. King, "Ion-Containing Polymers", Academic Press, New York, 1977.
3. W. J. MacKnight and T. R. Earnest, Jr., Macromol. Revs., 16, 41 (1981).
4. R. D. Lundberg and H. S. Makowski, "Ions in Polymers", Ed. A. Eisenberg, Adv. Chem. Ser., No. 187, ACS, Washington, D.C., 1980, Chapter 2.

5. M. F. Hoover and G. B. Butler, J. Polym. Sci., Polym. Symp., 45, 1 (1974).
6. K. Sanui, R. W. Lenz and W. J. MacKnight, J. Polym. Sci., Polym. Chem., 12, 1965 (1974).
7. C. Azuma and W. J. MacKnight, J. Polym. Sci., Polym. Chem., 15, 547 (1977).
8. H. S. Makowski, R. D. Lundberg, L. Westerman and J. Bock, "Ions in Polymers", Ed. A. Eisenberg, Adv. Chem. Ser., No. 187, ACS Washington, D.C., 1980, Chapter 1.
9. B. C. Johnson, I.Yilgor, C. Tran, M. Iqbal, J. P. Wightman, D. R. Lloyd and J. E. McGrath, J. Polym. Sci., Polym. Chem., 22(3), 721 (1984).
10. J. P. Quentin, Sulfonated Polyarylether Sulfones, U.S. Pat., 3,709,841, Rhone-Poulenc, January 9, 1973.
11. A. Noshay and L. M. Robeson, J. Appl. Polym. Sci., 20, 1885 (1976).
12. R. D. Allen, T. L. Huang, D. K. Mohanty, S. S. Huang, H. D. Qin and J. E. McGrath, Polym. Prepr., 24(2), 41 (1983).
13. R. A. Weiss, R. D. Lundberg and A. Werner, J. Polym. Sci., Polym. Chem., 18, 3427 (1980).
14. B. Siadat, B. Oster and R. W. Lenz, J. Appl. Polym. Sci., 26, 1027 (1981).
15. J. Selb and Y. Gallot, Polymer, 20, 1259 (1979).
16. J. Selb and Y. Gallot, Polymer, 20, 1273 (1979).
17. J. Selb and Y. Gallot, Makromol. Chem., 181, 809 (1980).
18. G. Broze, R. Jerome and Ph. Teyssie, Macromolecules, 14, 224 (1981).
19. G. Broze, R. Jerome and Ph. Teyssie, Macromolecules, 15, 920 (1982).
20. G. Broze, R. Jerome, Ph. Teyssie and C. Marco, J. Polym. Sci., Polym. Phys., 21, 2205 (1983).
21. S. Bagrodia, Y. Mohajer, G. L. Wilkes, R. F. Storey and J. P. Kennedy, Polym. Bull., 9, 174 (1983).
22. A. Noshay and J. E. McGrath, "Block Copolymers: Overview and Critical Survey", Academic Press, New York, 1977.
23. J. E. McGrath, Ed., "Anionic Polymerization: Kinetics, Mechanisms and Synthesis" ACS Symp. Ser., No. 166, Washington D.C., 1981.
24. M. Morton, "Anionic Polymerization: Principles and Practice", Academic Press, New York, 1983.
25. R. D. Allen and J. E. McGrath, Polym. Prepr., 25(2), 9 (1984).
26. I. Yilgor and J. E. McGrath, to be published.
27. Z. Wojtzack and K. Suchocka-Galas, IUPAC Int. Symp. Macromol., Strassbourg, France (1981). Proc. p. 311.
28. I. Yilgor, K. A. Packard, J. Eberle, R. D. Lundberg and J. E. McGrath, IUPAC Int. Symp. on Macromol., Bucharest, Romania (1983) Proc., I-10, p. 27.
29. N. Kornblum and S. Singaram, J. Org. Chem., 44, 4727 (1979).
30. S. Gauthier and A. Eisenberg, Polym. Prepr., 25(2), 113 (1984).

RECEIVED October 23, 1985

7

Relaxation in Styrene–Methacrylic Acid, Sodium Salt Ionomers

M. H. Litt[1] and G. Bazuin[2]

[1]Department of Macromolecular Science, Case Western Reserve University, Cleveland, OH 44106

[2]Department of Chemistry, McGill University, Montreal, Quebec, Canada H3A 2K6

The relaxation of styrene/Na methacrylate ionomer copolymers is studied in this paper. The present approach assumes a Rouse-Bueche distribution of relaxation strengths, with relaxation times defined by the number of ionic groups in a given segment (and the distance between the ions when a segment has two ionic groups). Good fits are obtained by curve fitting for all the polymers with ionic contents between 0 and 5.5 mole %. This is the range where time-temperature superposition holds. Parameters are reasonably consistent for all the runs and are physically plausible.

While ionomers of many types have been made and characterized [1,2,3], there is little work on the overall relaxation mechanisms. For polymers with low ionic concentrations, there is general agreement on the fundamental relaxation step. The stress relaxes by detachment of an ion pair from one cluster and reattachment to another. For the styrene/methacrylic acid Na salt (ST/-MAA-Na) system, there is a secondary plateau in the relaxation modulus which depends on the ionic content and can be described as a rubbery modulus [4]. While a rubbery modulus with stress relaxation due to ionic interchange has been invoked earlier, it does not adequately describe the relaxation curves. A different approach is taken here.

We assume that the basic relaxation strengths in the polystyrene relaxation spectrum as described by Ferry and others [5], based on a Rouse-Bueche [6] model, are maintained in the ionomer. A normal wedge-box distribution is assumed. The ions are in multiplets below 6 mole % ionic groups [2]. If a segment

0097–6156/86/0302–0093$06.00/0

contains no ionic groups, it relaxes exactly as it would in polystyrene. However ionic groups pin the segments. If a single ionic group is within a segment, it retards the relaxation by some factor, but the residues can relax around it. If a segment contains two or more ionic groups, it is pinned and cannot relax until the ionic groups relax by their normal exchange process. This is usually much slower than segmental relaxation, so the non-ionic segmental relaxation can be neglected. By analogy with the Rouse-Bueche model, the relaxation rate constants of a segment containing n ions is K/n^2. Such an assumption is also justified by the relaxation behavior of ionomers of various degrees of neutralization [7]. The model detailed below was applied to the data of Navratil in order to quantify it [4,8].

Nomenclature

- a = Mole fraction of anions
- b = Retardation factor for relaxation of segment with single ion
- E_o = Maximum modulus in relaxation spectrum
- E_N = Relaxation modulus of segment with N residues
  $E_N \cong E_o/N^2$, $N \leq 300$
- E_f = Plateau (box) relaxation modulus, $N > 300$
- K = Rate constant of single ion-pair exchange relative to τ
- K_n = K/n^2 = Relaxation rate constant for segment containing n ion pairs, $N \leq 300$. $K_n = 300aK/n^3$, $N > 300$, $n > 300a$
- L = Length of "freely rotating segment"
- N = Number of residues per segment
- N_f = Number of residues per molecule
- n = Number of ion pairs per segment
- τ = 1 sec = Normalized relaxation time of segment with one residue
- τ_N = $\tau \cdot N^2$ = Relaxation time of segment with N residues
- Z = Average distance between ions in bulk system before coupling.

Theory

The fundamental equation is given below and its various aspects are then discussed. It was applied to data on styrene/methacrylic acid copolymers, Na salts of Eisenberg and Navratil [4,8].

$$E = \sum_{N=1}^{N_f} E_N[(1-a)^N \exp(-t/\tau_N) + N(1-a)^{N-1} a \exp(-t/b\tau_N)$$

$$+ \sum_{n=2}^{N} \frac{N!}{(N-n)!n!} (1-a)^{N-n} a^n \exp(-K_n t)] \quad (1)$$

As is normal in such cases, N is stepped exponentially. Values of N taken are log linear. We have used a factor of $10^{0.25}$ for most of the work. In order to simplify minimization, E was described as a continuous function.

$$E_N = E_o/N^2 + E_f \quad (2)$$

E_o is about $10^{9.5}$ Pa and E_f is about 10^5 Pa. Since the degree of polymerization was not known for Navratil's polymers, there was no reason to change the form of τ_N for large segments. The effect of keeping $\tau_N = \tau/N^2$ is simply to add several more terms to the summation with a corresponding drop in E_f.

If we assume a random distribution of ionic groups along the polymer chain (a doubtful assumption, but no other is possible at this approximation level), then for segments of length N, with an ionic fraction a, the fractions of such segments with no, one, two, etc. ionic groups is described by a binomial expansion, Equation 3.

$$[(1-a) + a]^N = (1-a)^N + N(1-a)^{N-1} a + \sum_{n=2}^{N} \frac{N!}{(N-n)!n!} (1-a)^{N-n} a^n \quad (3)$$

Each term is the weighting factor multiplying the corresponding relaxation in Equation 1. Segments without an ionic group relax at the rate at which a similar segment from pure polystyrene relaxes. The retardation of relaxation of a segment when it contains one ionic group is intuitively plausible. As one residue in the segment is pinned, the other residues must relax around it. Relaxation is more difficult than when all segments can participate in the process.

The logic applied to segments with one ion-pair also holds for small segments with two ion-pairs. If the ion-pairs are too close along the backbone to bind into separate multiplets, they must associate in a single multiplet. Relaxation should be strongly retarded compared to similar segments with no ion-pairs, but need not depend on ion-pair dissociation to facilitate relaxation. We have assumed arbitrarily that the retardation factor for a short segment containing two ion-pairs is the square of the retardation factor of a single ion-pair containing segment. The number of residues needed in a segment before its ion pairs are in separate multiplets will vary with the ion concentration. It was left as an adjustable parameter, the cut-off value. In a homogeneous system with Gaussian statistics along the chain, the segment length at the cut-off point should be

$$NL \approx 2(60/a)^{2/3}/L = \text{Segment Length} \quad (4)$$

where L is the length of a freely rotating segment. This will be considered further in the Discussion section.

The same logic could be applied to segments containing three or more ionic groups where two are coupled together. The relaxation times should be different from those segments where all ionic groups are paired with groups from different segments. However, this is too complicated to consider here.

When segments have two or more ionic groups, and these are bound in separate multiplets, the segments cannot relax except by multiplet relaxation. We have assumed that the relaxation rate per segment for $N \leq 300$ depends inversely on the square of the number of ions. In part this is by analogy with the dependence of τ_N. When $N \geq 300$, the relaxation rate has been taken as proportional to n^{-3}. One report justifies this assumption [7]. The report shows the effect of changing neutralization on the longest relaxation time of an ethylene/methacrylic acid (5.4 mole % acid) copolymer. The data are inexact, but show an increase in τ from 10^3 to 10^5 sec as the neutralization goes from 5 or 10% to 30%. This is a 2.5 to 4 power dependence of τ on the ion concentration. Long segments seem to relax at about the inverse third power of the ionic concentration. This could be considered the equivalent of the longest τ for a normal polymer which depends on $(\bar{M}_w)^{3.4}$. Again, since molecular weights were not determined for the data considered here, the choice of an exponent for n is almost arbitrary. We have chosen n^3 when $N > 300$ to save time in the summations. A smaller exponent can be compensated by a larger value of N_f and vice versa. [Even at very low ionic constant, all large segments contain many ions and the pure polymer rubbery relaxation vanishes. We need only consider ionic relaxation.]

Application to Data

The relaxation data studied were confined to $a < 6\%$, where time-temperature superposition was valid [2,4,8]. In the present approach, the time shifted data were used, as less parameters are needed. The important parameters that must be fitted are E_o, E_f, b and K. Other parameters must be determined in order to fit the theory to the data, but they are not intrinsic to the system. They are:

1. The time shift value: This moves the curve along the log time axis, which is equivalent to finding a value for τ. (When a time shift is used and τ is kept as 1, K becomes the ratio of the ionic to segmental relaxation rate constants.)
2. The number of terms used in the summation: This determines N and is fixed by the "time" at which viscous flow starts. While the tail of the relaxation curve is adjusted by this parameter, it is a cosmetic fit since we do not know the molecular weight, and the real polymer is polydisperse.

The central region of the curve, comprising the drop to the ionic plateau, the plateau modulus, and its drop-off towards the rubbery plateau, is determined by the form of Equation 1, K, E_o, E_f and b.

The data points and the minimized curves based on Equation 1 for the six styrene/MAA-Na copolymers with ionic mole fractions ranging from .006 to 0.055 are given in Figures 1a and 1b. A polystyrene relaxation curve is also shown. The values of the parameters used are listed in Table 1, together with the Tg's of the copolymers and the standard error of the fit.

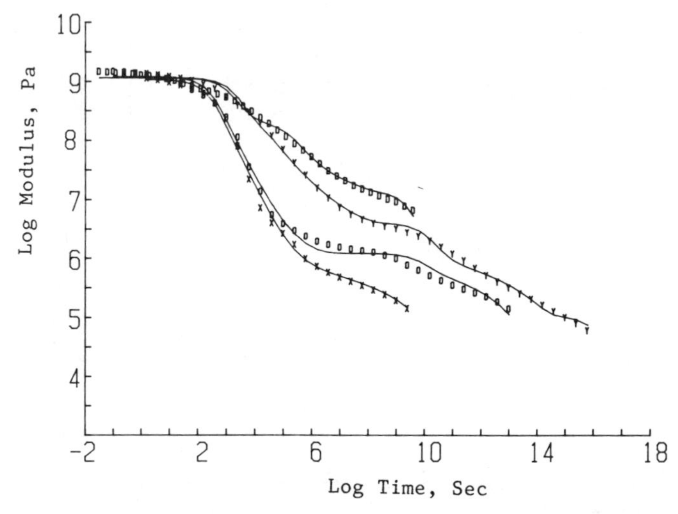

Figure 1a. Comparison of Equation 1 (-) using values in Table 1 with experimental points from the relaxation of NAV 000 (X), NAV 019 (O), NAV 038 (Y), and NAV 055 (D).

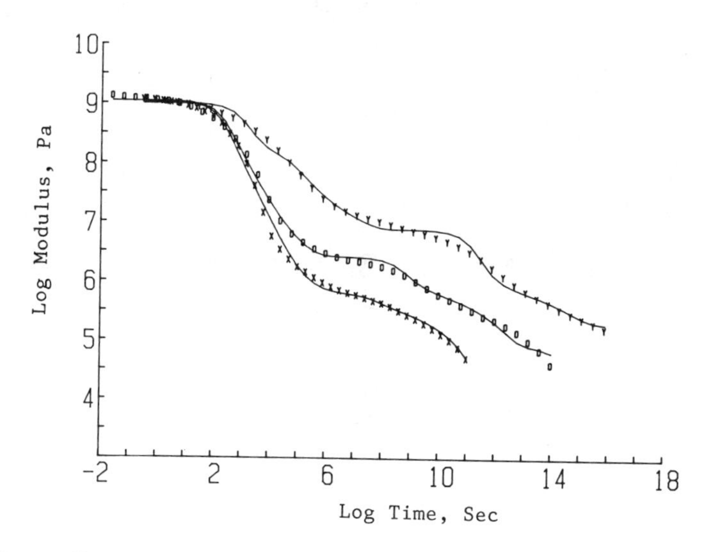

Figure 1b. Comparison of Equation 1 (-) using values in Table 1 with experimental points for the relaxation of NAV 006 (X), NAV 025 (O) and NAV 046 (Y).

Table 1. Values of Parameters Used in Equation 1 for Best Fit.

Sample	Ion Content	log Δt	Log E_0, Pa	log E_f, Pa	log K	No. of Terms	SIRF[b]	Cut-Off	S.E.$(X10^2)$	Tg, °C
NAV 000[a]	0	-1.710	9.573	4.848	-	17	-	-	7.1	102
NAV 006	0.006	-1.542	9.572	4.759	-5.551	17	1.02	75	7.6	105
NAV 019	0.019	-1.740	9.634	4.911	-7.301	16	2.51	75	9.7	109
NAV 025	0.025	-1.602	9.580	5.020	-6.321	16	3.72	30	7.8	110
NAV 038	0.038	-2.562	9.604	5.036	-6.716	16	22.80	30	7.0	112
NAV 046	0.046	-2.285	9.541	5.041	-8.198	16	69.34	12	8.0	114
NAV 055	0.055	-2.375	9.602	5.010	-6.461	12	139.6	7	7.7	110

[a] The samples studied by Navratil (4) have been relabeled so that the numbers in the label reflect the ionic content.

[b] Single Ion Retardation Factor

Discussion

It is obvious from Figure 1 and Table 1 that Equation 1 fits the data reasonably well. The standard error is about .08 or $\pm$ 20% for any value of the relaxation modulus. Since the modulus covers over four decades, this is reasonable. The values for E_o and E_f remain almost constant for all the polymers. In this approach, log (Δ t) should be constant if the data were normalized to a given temperature. However, the data were normalized to log(time) equals 0 at T_g. Thus there is a gradual shift of log (Δ t) as ionic content increases, to compensate for the shift of the reference temperature. However, the maximum shift in log(Δ t) is about 1 (between NAV 006 and NAV 038), while the shift of log a_T (compared at a constant reference temperature) is about 2.48. It may be possible that segmental relaxation is hindered by the presence of ionic groups in neighboring segments, thus raising the temperature for the approach to T_g, though the presence of ionic groups is the major cause of the rise in T_g. A similar decrease in segmental relaxation rate as a function of ion concentration can be seen in the Single Ion Retardation Factor. This is 1.0 for NAV 006 and rises to 140 for NAV 055. The fact that the non-ion bearing segmental relaxation rates are independent of ionic concentration in the present approach produces some of the discrepancy between theory and experiment. The theoretical curves for NAV 046 and NAV 055 show a rapid relaxation at the start of the drop in log modulus which is not reflected in the experimental data. The experimental curve shows a smooth drop with the slope varying inversely with the ionic content.

The position of the first plateau, at log time $\approx$ 6 depends sensitively on the cut-off chosen. This is one factor which affects the fit between theory and experiment very much. However, the resolution is too coarse; the factor by which the segment's length changes, $10^{0.25}$ (=1.78), is such that some of the curves cannot be fit exactly. However, doubling the resolution would increase computation time enormously.

One can make a rough estimate of the average segment length between ions at which cross-over between intermolecular and intramolecular coupling occurs. Consider the following "thought experiment". The polymer with ions on it is placed in an equilibrium conformation, without allowing the ions to couple. The ions are then distributed randomly in space. When they are allowed to couple, they couple with their nearest neighbor. If the random walk distance of ions along the chain is less than the average interionic distance, intramolecular coupling will occur. Otherwise, intermolecular coupling will occur. In terms of a, the mole fraction of ions in polystyrene, considering the residue molarity in bulk polystyrene to be 10M, there are .01a moles of ions per cubic centimeter. Thus the average distance between ions in nm is:

$$Z = (.167/a)^{1/3} \qquad (5)$$

If we take L = .7 nm (9), then N can be calculated, since NL = 0.2 nm/a. Thus $N L^{1/2}$ can be found easily. The change from intramolecular coupling to intermolecular coupling occurs when the segment is $\geq$ 2Z, since on the average two ionic groups in a segment will be a half segment length apart. Thus, the change-over condition is given in (6),

$$N^{1/2}L = Z \qquad (6)$$

with the segment length in Å equal to 2NL. One squares (6) and solves for NL to obtain (4). Thus we find: No. of residues at cut-off = 2NL/2.5 (7). The data are given in Table 2 and compared with the cut-off values found from the minimization. The cut-off values are arbitrary within the range defined by the successive values of N chosen. These are listed in the last column of Table 2. One can see that up to a = 0.038, the fit is very good. The plateau modulus rises faster than this approach predicts for the higher a values, thus leading to lower cut-off values. This may be due to the beginning of cluster formation, which dominates the relaxations above a = 0.06.

The assumptions made for Equation 4 are drastic. We assume that the material behaves like pure polystyrene during the molding and that all chains are in equilibrium positions. Then all ions interact with those nearest to them to generate the multiplets. This approach is a zero order approximation. Ionic association is certainly history dependent. Only ion pair dimers are considered, not higher multiplets. However, NL agrees with the cut-off values at the lower ion concentrations, which strengthens the overall hypothesis.

The number of terms are reasonable. A value of 16 corresponds to 10^4 mers per chain of $\bar{M}w \approx 10^6$. Navratil measured the number average molecular weights of two of the samples discussed here (8) and found $\bar{M}n$(NAV 019) = 470,000 and $\bar{M}n$(NAV 038) = 400,000. Thus, the value of N_f found is in good agreement with the experimental. Sample NAV 055 is listed as a low molecular weight polymer (8) and we find that only 12 terms are necessary in the summation. Thus, $\bar{M}w \approx$ 100,000 for NAV 055 based on the curve fitting. All the other polymers were "high molecular weight" according to Navratil. This agreement implies that the assumption $K_n \alpha 1/n^3$ for large segments is reasonable. An exponent of 3.5 or 4 might also work.

Table 2. Comparison of Theoretical and "Experimental" Cut-Off Values

	No. of Residues in Segment at Cut-Off		
a	Theory	"Experiment"	Cut-Off Range
.006	104	75	46-81
.019	48	75	
.025	40	30	26-46
.038	30	30	
.046	27	12	8-14
.055	24	7	4.6—8

The standard error is measurable. While the modulus/time plots are smooth for a given temperature, we are joining many such curves together to get the master curve. Navratil has shown (8, p. A2) that when the modulus is not changing rapidly, reproducibility of a given segment of the master curve is excellent ($\geq$ 10%). However, the curves diverge the most at very short and very long times which is what is used when they are fitted to make the master curve. Thus a standard error in log modulus of .08 ($\pm$ 20%) is quite good.

The values of log K vary randomly in the range of -6.8 $\pm$ 1.3 for the different samples. Two values are probably incorrect, those for NAV 019 and NAV 046, a consequence of curve fitting for the lowest standard error. This can be seen easily in inspecting Figures 1a and 1b. The rest cluster around log K = -6.0 WU$\pm$ .45. Any trend due to increasing ionic content (which might lower K) or increasing normalization temperature with increasing ionic content (which would tend to raise K) is too small to be distinguished.

Log (Δ t) - log K $\approx$ 4.1 for three of the four remaining curves and is 4.56 for the fourth. There is possibly a constant displacement between the two terms, implying that the ionic dissociation rate constant is about 10^4 smaller than $1/\tau$.

The sources of the discrepancies between the experimental data and the theoretical curves need some discussion. First, the model assumes a monodisperse polymer. This affects the curve shape after the upper ionic plateau. Second, the model assumes that all ions (except for a few very close together) are coupled into multiplets with ions from other chains. There are few loops. The analysis of Equation 4 suggests that perhaps most of the ions are coupled to adjacent ions in the same chain. This would change the nature of the relaxations considerably and affect the modulus at the ionic plateau. Such a model could lead to semi-permanent entanglements, kep in place by the ionic multiplets. The retardation of all relaxations of small segments at higher ionic content, even that fraction which occurs in ion-free segments is not considered in this model.

Conclusions

A model for relaxations in ion containing polymers was presented here. It fits the experimental data for St/MAA-Na copolymers reasonably well. The parameters which are obtained from the fit are reasonable, though there is uncertainty in the exact values. One further result of this analysis is the realization that all relaxations are retarded even in the high modulus portion of the relaxation curve. This suggests that an analysis of coupled chain oscillations for this system might be in order.

Literature Cited

1. See Ionic Polymers, ed. L. Holliday, Applied Science Publishers, London, 1975.
2. See A. Eisenberg and M. King, Ion Containing Polymers, Academic Press, New York, 1977.
3. A. Eisenberg, Macromolecules, 4, 125, (1971).
4. M.F. Hoover, J. Polymer Sci., Polymer Symposium 45, 1, (1974).

5. A. Eisenberg and M. Navratil. Macromolecules 6, 604, (1973).
6. See, for example J.D. Ferry, Viscoelastic Properties of Polymers, 3rd ed., John Wiley & Sons, New York, 1980, Chapter 2 and Appendix D. For a general discussion, see Polymer Science and Materials, ed. A.V. Tobolsky and H.F. Mark, John Wiley & Sons, New York, 1971. Chapter 10, by A.V. Tobolsky.
7. F. Buuche, Physical Properties of Polymers, Interscience, New York. 1962.
8. R. Longworth and D.J. Vaughn, unpublished results, quoted in Ref. 1, p. 108.
9. M. Navratil, Ph.D. thesis, Nov. 1972, Dept. of Chemistry, McGill University, Montreal, P.Q., Canada.
10. Polymer Handbook, 2nd ed., ed. J. Brandrup and E.H. Immergut, John Wiley & Sons, New York, 1975 pp. IV 40,41.

RECEIVED June 17, 1985

8

Clustering and Hydration in Ionomers

Bernard Dreyfus

Groupe de Physico-Chimie Moléculaire, Département de Recherche Fondamentale, Service de Physique, Centre d'Etudes Nucléaires de Grenoble, 85 X, 38041 Grenoble Cedex, France

A new model for the clustering of charges in dry ionomers is presented. The basic idea is that, under the influence of electrostatic interactions, the multiplets of charges coalesce in clusters that have an internal structure compatible with the steric hindrances due to the polymeric material. The size of the cluster is shown to be independent of the concentration of charges. The tension of the chains within the matrix is discussed, and it is suggested that the clusters are arranged in small hypercrystallites with a local order of the diamond type. The free energy of a "vesicle" containing the water molecules and the mobile cations is calculated. The cations are shown to build a layer of charges; at equilibrium the fixed anions coat the vesicle as densely as they can. Thus the ideal solution formula for the chemical potential of water is not valid. A few consequences are discussed.

Ionomers are copolymers containing a molar concentration c of charged neutralizable groups ($c < 0.1$). Their properties are dominated by the interplay between the charges and the chains. We exclude here the case of Nafions, where the charges are not close to the polymeric chains.

In the last fifteen years many experimental studies and several models have been devoted to these materials. Several reviews of this subject have been published (1-4). It is generally recognized that the charges condense into regions called clusters with a radius in the range 10 to 30 Å, containing typically 50 to 100 basic pairs of charges. The existence of a universal "ionomer peak" in scattering experiments has led to the idea that the clusters are separated from each other by distances D of the order of 20 to 80 Å.

The objective of this paper is to present a simple mathematical model for clustering in dry materials. We refer to a more detailed version of the calculation (5). The strongly hydrated state, when the water molecules and the mobile cations form kinds of vesicles (or inverted micelles) will be examined later in this paper. As in any model, our mathematical model is rather ideal in the sense that it

0097-6156/86/0302-0103$06.00/0

ignores many of the pecularities of the polymers and of the ions. As it will lead to easily tractable results, this model may be a guideline for the experimentalists.

In an early work, Eisenberg (6) showed that the condensation of charges is a two-step process. The basic ionic pair (one anion plus one cation) is closely associated with a few other pairs to form a multiplet. The electrostatic energy responsible for the formation of the multiplet is of the order of e^2/d (e = electronic charge, d the distance between the centers of the anion and of the cation). We call this energy the primary electrostatic energy; it is in the range of 100 kcal and it cross-links strongly different chains. The size of the multiplet is limited by steric hindrances; if k is the "functionality" (number of cross-linked chains), 2 k segments have to leave the multiplet. These "coated" multiplets still interact through the fields of their electrical multiplets (k = 2 is a quadrupole, etc.). We call this interaction the residual electrostatic interaction. It is still important: for two quadrupoles of moment $e\,d^2$ at a distance r, this interaction is on the order of $e^2 d^4/r^5$. Numerically for $d = 3$ Å, $r = 8$ Å and $k_BT = 0.6$ kcal, one gets $1.38\ k_BT$ in vacuum; 8 Å is the average distance between quadruplets in a sample with $c = 0.05$ and v = volume of one monomer = 50 Å^3. The attractive interactions cause a coalescence of the multiplets which is limited only by the hard-core repulsion between the monomers. In other terms, the density cannot exceed that of the neutral material. It will be shown that this basic fact leads to a cluster with a well-defined density of charges. It is only at the surface of the cluster that the tension of the chains, through their entropy, plays an important role and limits the size of the cluster.

The dry clusters

Structure of the Cluster. Definition of variables is important ; in the following discussions we use the following notations : c, molar concentration of the cluster; ρ, distance from its center; v, σ, ℓ, volume, cross-section, and length of one monomer. These values are obtained from crystallographic data. The chains are treated as ideal free jointed rods of monomers. For numerical applications, we generally use the case of polyethylene: $v = 50$ Å^3, $\sigma = 20$ Å^2, $\ell = 2{,}5$ Å. $(1 + \alpha)v$ will be taken as the volume of one neutralized charged ionomer. In many applications we shall take, for simplicity, $\alpha = 0$ for the dry state. If the cations are solvated by ν water molecules for each of volume v_0 (30 Å^3), a simple additivity rule for the volumes give $\alpha v = \nu v_0$.

The procedure by which we build the cluster is as follows: we call $n(\rho)$ the number of basic pairs already inside a sphere of radius ρ. Because of the presence of these pairs a certain number $s(\rho)$ of segments have to leave the sphere.

The space in the shell btween ρ and $\rho + d\rho$ is attributed to these segments plus, to fill it, a certain number of "new" charges $dn(\rho)$. The basic assumption is that the neutral segments are expelled from this deep ionic region as directly as possible; we assume they go out radially. Attention is paid to the fact that a small proportion (c) of these segments carry charges (Figure 1). We are led to a pair of differential equations in $n(\rho)$ and $s(\rho)$. The detail of the calculation is given in Ref. 5. We shall give here only a simple physical argument, approximate, but sufficient for what follows. The maximum dense packing of charges is realized when the total cross-section of the segments issuing from the sphere of radius ρ, is equal to the surface offered.

$$4\pi\rho^2 = 2\ \sigma\ n(\rho) \tag{1}$$

The complete calculation (5) gives an exact formula

$$n(\rho) = \frac{2\pi\ell}{kv(1+\alpha c)} \left[\frac{2c}{3\ell}\rho^3 + \rho^2 \frac{1-c}{1+\alpha c} - \rho\ell \frac{(1+\alpha)(1-c)}{(1+\alpha c)^2} + \frac{\ell^2}{2} \frac{(1+\alpha)^2(1-c)}{(1+\alpha c)^3} \left(1 - \exp - \frac{2\rho}{\ell}\frac{1+\alpha c}{1+\alpha}\right) \right] \tag{2}$$

from which the volume density of charge

$$g(\rho) = \frac{dn(\rho)}{4\pi\rho^2 d\rho} \tag{3}$$

can be obtained. If we take into account that c is a small number and $1 + \alpha c$ is close to 1, one has :

$$vg(\rho) = c + \frac{\ell}{\rho} - \frac{\rho^2}{2\rho^2}(1+\alpha)\left(1 - \exp - \frac{2\ell}{\ell(1+\alpha)}\right) \tag{4}$$

This function is plotted in Figure 2, for $\alpha = 0$, and $\alpha = 2$. The dashed line represents the approximation obtained from Equation 1 :

$$vg(\rho) = \frac{\ell}{\rho} \tag{5}$$

This approximation is sufficient in the useful range of ρ. The fact that it is poor in the very center of the cluster is unimportant because it is weighted by the geometrical factor $4\eta\rho^2$. In the case of a solvated sample ($\alpha \neq 0$) the next significant approximation would be :

$$vg(\rho) = \frac{\ell}{\ell} - \frac{\ell^2}{2\rho^2}(1 + \alpha) \tag{6}$$

The conclusion of this section is that the dense packing of the material, combined with the attraction of the multiplets, leads to a structure of the cluster described by a law in ρ^{-1} (it must be noted that this law is independant of k).

The Radius ρ_c of the Clusters. We still have to look for what limits the size of the cluster. Until now we have neglected the role of the matrix in which the cluster is embedded. When a segment of ν monomers is free, its natural end-to-end length is

$$\bar{R}^2 = \nu\ell^2 \tag{7}$$

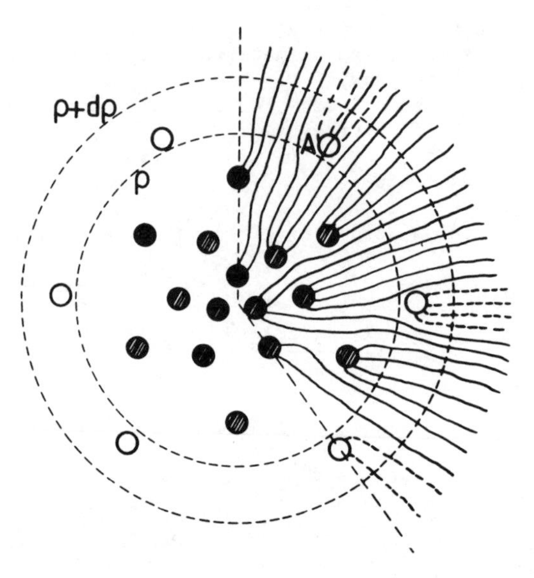

Figure 1. A two-dimensional view of a cluster. Cross-hatched circles are multiplets inside a circle of radius ρ. Four segments are shown going out from each multiplet, and are represented by continuous lines. Open circles are the new multiplets in the shell between ρ, ρ + dρ. The "new" segments are represented by broken lines. At point A a "new" multiplet is located on a segment issuing from a more internal multiplet. (Reproduced from Ref. 5. Copyright 1985 American Chemical Society.)

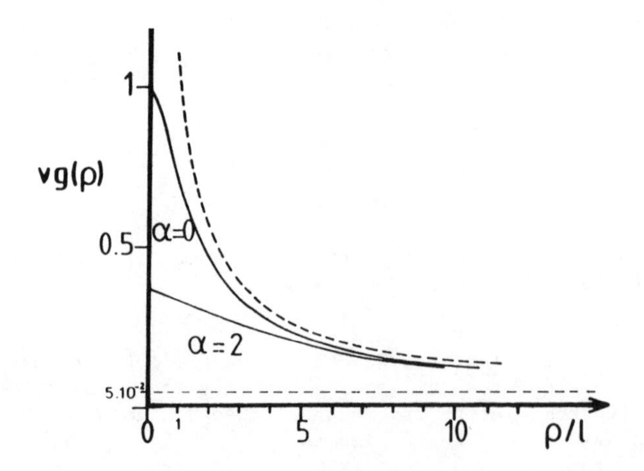

Figure 2. Density vg(ρ) as a function of ρ/ℓ, for $c = 5\ 10^{-2}$, $\alpha = 0$ and $\alpha = 2$. The dashed line represents the approximation $vg = \frac{\ell}{\rho}$. Note that solvation affects mainly the cluster's core. (Reproduced from Ref. 5. Copyright 1985 American Chemical Society.)

If one imposes a distance R between the two ends of the segment, for a variation $\Delta(R^2)$, its entropy varies like :

$$\Delta s = -\frac{3}{2} k \frac{\Delta(R^2)}{R^2} \quad (8)$$

$\Delta(R^2)$ is expected to be of the order of R^2; for example if the final end-to-end distance is R/2 (the multiplet relaxes toward the middle of two clusters) then $\Delta(R^2) = 3/4\ R^2$. For R^2 itself of the order of a few units of R^2, one sees that the absoption of one multiplet from the matrix to the surface of a cluster, corresponds to an entropy variation of the order of $- k_B$. We determine the radius ρ_c of the cluster by saying that the variation of the free energy for the surface multiplets is zero at equilibrium. The variation of the electrostatic energy is given by its value at the surface, since in the matrix the multiplets are far from each other.

An order of magnitude for the electrostatic energy of these surface multiplets is obtained (7) from

$$W = \frac{e^2}{\varepsilon' d} \left(\frac{d}{u} \right)^{2k+1} \quad (9)$$

where ε' is the dielectric constant of the matrix(usually on the order of 2) and u the distance between two k-plets, which can be obtained from their density by Equation 5.

Making W and k_BT equal, the expression for the radius becomes:

$$\frac{\rho c}{\ell} = \left(\frac{e^2}{\varepsilon' d} / k_BT \right)^{\frac{3}{2k+1}} \frac{d^3}{kv} \quad (10)$$

where d is the length of the elementary dipole.

Of course, the approximations on S and W (Eq. 8 and 9) are rather crude; however, because W is a rapidly varying function of u, a more exact approximation could not affect ρ_c too strongly.

For polyethylene, and for d : 3 Å and 4.5 Å, the radius ρ_c and number of charge are calculated for several values of k (see Table I).

Table I. Calculated Radius and Number of Charge for Polyethylene.

	d = 3 Å			d = 4.5 Å		
k	1	2	4	1	2	4
ρ_c (Å)	125	11	1,5	282	28	4,5
n_c	4960	33	0,75	25100	231	6,5

It is clear from this table that k=2 gives realistic values for ρ_c and n_c. As all the previous calculations are without any ajustable parameter, these calculations provide an argument in favor of quadruplets, i.e, a dimerization of the charged units. The scaling law should be noted:

$$\rho_c \sim d^{2.4}\ \sigma^{-1} \qquad n_c \sim d^{4.8}\ \sigma^{-3} \tag{11}$$

These relationships show the importance of d. When the elementary ions are large (weak salts) the size (ρ_c) and capacity (n_c) of the clusters become large. The radius ρ_c can be expressed as a function of a limiting concentration c_o at the surface of the cluster:

$$\rho_c = \ell / c_o \qquad n_c = \frac{2\pi\ell^2}{\sigma c_o^2} \tag{12}$$

In practice c_o can be considered an ajustable parameter. It increases for weak electrostatic interactions.

The Matrix. The condensation of quadruplets in clusters implies displacement of the charges and tension in the matrix that we shall try to estimate. We have first to define the spatial arrangment of the clusters. If we think that the clusters make a crystal, we must look for a structure, which for a given density of lattice points gives the shortest distance D between first neighbors in order to minimize the tension. This minimization is realized by the diamond structure (Z : number of 1st neighbors = 4).

Assuming that all the charges condense in clusters, the conservation of their number leads to a relation between n_c and D:

$$n_c = (8/3\sqrt{3})\frac{c\ D^3}{v} = 1.54\ c\ D^3/v \tag{13}$$

$$D = 1.60\ (\rho_c^2 \ell)^{1/3}\ c^{-1/3} \tag{14}$$

If we identify D with the intercluster distance deduced from the ionomer peak, this $c^{-1/3}$ law has been observed in butadiene-methacrylic ionomers (8) and pentanemer sulfonate (9). Conversely, a law of this type cannot be explained in the absence of clusters of capacity independant of c, such the one described in this model. When cations are substituted, resulting in a change of d, D should change, at constant c, as: $D \sim d^{1.6}$ (15)

Observations in neutralization experiments (9-11) are in agreement at least qualitatively, with Equation 15. The biggest cations give the largest D (if one accepts that the proton of the acidic form is the smallest cation).

Once a value is determined for D, one may estimate the state of tension of the segments in the matrix. The average number of monomers in these segments is 1/c , and the corresponding free end-to-end length is

$$\bar{R} = c^{-1/2}/\ell \tag{16}$$

With all the above formulas we can calculate the ratio $D/\bar{R}$, which gives an idea of the tension of the segments connecting two clusters. We also calculate the ratio $(D - 2\rho_c)/\bar{R}$, which is valid for the shortest distance between two clusters as, as we shall see, $2\rho_c$ is not negligible compared to D. In Figure 3, the results are plotted as a function of c, for two values of c_o (0.1 and 0.3).

$D/\bar{R}$ increases slowly with $c^{1/6}$. $(D - 2\rho_c)/\bar{R}$ is rather flat. The decrease at large c reflects that for a concentration $c = 0.51\ c_o$, the spherical clusters actually contact. In that region we expect that the clusters will fuse together and make a three-dimensional lattice of channels through the sample. Figure 3 shows that the segments are extended in a moderate way; the average tension may be considered as staying between the two curves.

There are several ways in which the material can use its internal degrees of freedom to decrease this tension (5). One of them implies that a fraction of the segments do not connect two adjacent clusters, but come back to the originating cluster. From the assumption that, before clustering, the chains make a three-dimensional unstrained network, and that each quadruplet will coalesce on the nearest cluster formed, one can obtain an expression for the probability P that a segment will be non-connecting (or will close back on the originating cluster). An exact calculation is given in the appendix of reference (5) ; a good approximation is given by:

$$P = 1 - \frac{1.20}{u^{1/2}} \qquad u = \frac{3\delta^2}{p\ell^2} \tag{17}$$

where δ is the radius of the sphere occupied by one cluster; p is the number of neutral monomers between 2 consecutive charges. Numerically, for $c_o = 0.1$ and $c = 10^{-2}$ or 10^{-3}, one gets $P = 0.74$ or 0.6. More than one half of the segments close back on the originating cluster. This fraction increases with c, and decreases with c_o. The shortest segments (small p) have a greater probability of returning.

These results show that when D can be varied either by substitution of cations or by hydration, the rearrangment of the order is a delicate process in which not only the tension of the segments is modified but also the proportion of connecting segments.

Scattering and Solvation. In X-ray or neutron diffraction experiments the scattered intensity along the wave vector Q ($Q = \frac{4\pi}{\lambda} \sin\frac{\theta}{2}$) is given by:

$$I(Q) = K\ F(Q)^2 S(Q) \tag{18}$$

F(Q) is the structure factor of an individual cluster, S(Q) an interference term. If the scattering is mainly due to the cations, either because of their atomic number or because of their solvation shells, F(Q) can be calculated from their density (Equation 5):

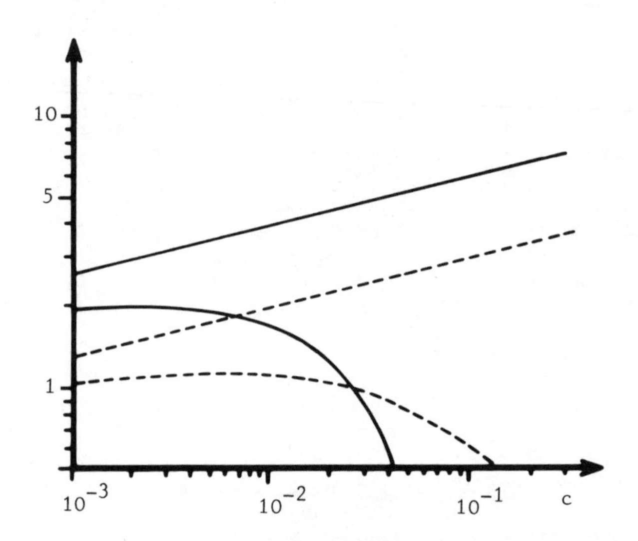

Figure 3. Plots of $D/\bar{R}$ (straight lines) and $(D - 2\rho_c)/\bar{R}$ (curves) as a function of c. Full lines $c_0 = 0.1$; dashed lines $c_0 = 0.3$. (Reproduced from Ref. 5. Copyright 1985 American Chemical Society.)

$$F(Q) \sim n_c \left(\frac{\sin x/2}{x/2}\right)^2 \qquad x = Q\rho_c \qquad (19)$$

For a constant density sphere, a different function results:

$$F(Q) \sim n_c \; 3 \, \frac{\sin x - x \cos x}{x^3}$$

To calculate S(Q), a model must be used for the spatial order of the clusters. Usually one assumes a random gas of clusters with a hard core potential of radius D. This method allows a reasonable analysis of the ionomers peak (11). We propose a different approach based upon the existence of a local order of the clusters. Although the distance D is so large that no electrical potential can have a significant influence on their positions, the clusters are linked by many segments that cannot be extended easily to accomodate large changes in D. Even the non-connecting segments yield a kind of elastic ball which coats the cluster. The end result is an "entropic interaction" which tends to hold the cluster on a regular lattice (hypercrystal). We have already suggested a diamond type lattice for the clusters.

Of course the range of this order cannot be very long, and we propose hypercrystallites of size Δ (Δ > D) of clusters. Thus as $F^2(QD)$ decreases rapidly, a possible explanation is offered for both an "ionomer peak" (with possibly some weak satellite at larger Q) and a central peak of width Δ^{-1}. Δ(in the range of 100 Å or more) could be due to the sample preparation or to the finite length of the polymeric chains, as $\sqrt{N}\,\ell$ is the only parameter of the problem in this range for N ∿ several thousands.

It is appropriate to end this section with a discussion of solvation. There are experimental indications that D tends to increase. The present model can help to distinguish several possible effects.

1 - The macroscopic expansion of the sample by a factor $(1 + \alpha c)$ if n_c is held constant. This is a weak term; for polyethylene and $\nu = 4$ water molecules per cation, this leaves an increase in D of 3.8%.

2 - The "inert" increase of volume of the charged monomers by a factor $1 + \alpha$ (3.4 for $\nu = 4$). As shown in Figure 2, this gives a strong decrease in the density of the charges, mainly in the core of the cluster. For the same electrostatic interaction (Equation 9) this gives a decreases in n_c of 48% for $\nu = 4$ and a decrease in D of about 13%.

3 - The most important effect probably lies in the weakening of the bond between anions and cations. We lack an expression for $d(\nu)$; as $n_c \sim d^{4.8}$ (Equation 11) it is probably the dominant one. If we does not take into account the preceding effects, an increase

in d only 15% would a double n_c and increase D by 26%. If we come back to the scaling laws

$$\rho_c \sim d^{2.4} \qquad n_c \sim d^{4.8} \qquad D \sim d^{1.6} \tag{20}$$

it follows that ρ_c increases more rapidly than D.

In the discussion of the matrix we noted that the dry clusters have physical contact when c is large. By solvation we expect that this will happen for a smaller value of c. One sees here the strong tendency of the clusters, favored by solvation, to collapse into a network of channels. Similar structures have been already proposed (13); they are of fundamental importance, for practical use, in explaining the striking transport properties of ionomers.

Strong hydration

In this part of the paper we examine the thermodynamic properties of hydrated ionomers. By strongly hydrated we mean that we are beyond the state of solvation shells, where ν, the number of water molecules per cation, is a small number ($\sim$ 4 to 6). In a strongly hydrated sample, the water molecules are considered to be free, and make a concentrated solution with the cations (the counter ions) and eventually with some mobile anions (the coions). This subject has already been extensively studied because of its practical importance (1-4). From the following discussion, we shall see that some of the usual classical laws are no longer valid. For instance, the variation of the chemical potential of water with the concentration of cations may no longer hold.

Electrostatic and Entropic Terms. Just as charges coalesce into clusters in the dry state, upon hydration, the charges and water molecules are gathered into vesicles, which we assume to be spherical (radius R). Experiments show that the location of the ionomer peak varies continuously with the water content (ν) Thus we assume that the sample is in thermodynamical equilibrium, and we have to look for its free energy.

We call N the number of fixed anions coating the vesicle. Here we consider only the case of electrically neutral vesicles, with no coions, so that N is also the number of mobile cations inside the vesicle. We introduce a reduced radius $x = \rho/R$, where ρ is the distance from the center. $\Gamma(x)$ is the distribution function of the cations; $\Gamma(x)\,dx$ is the probability of finding a cation between x and x + dx. The volume density probability $\gamma(x)$ is related to $\Gamma(x)$ by

$$\Gamma(x) = 4\Pi \; x^2 \; \gamma(x) \tag{21}$$

The shape of $\Gamma(x)$ will be determined by minimizing the free energy G, which is a function of $\Gamma(x)$. In order to calculate G we need an expression for the electrostatic energy and for the entropy of the system. The complete details of the calculation of the electrostatic energy will be given in a forthcoming paper (14).

Water and matrix are characterized by their respective dielectric constants, ε and ε'. Typical value are $\varepsilon \sim 80$, $\varepsilon' \sim 2$, so that $\varepsilon' << \varepsilon$. We first write the self energy of one cation in the vesicle. It expresses the repulsion by the electrical image due to the spherical boundary. As is well known (15), there is no exact analytical expression for it in the spherical geometry, but if we neglect terms in $1/\varepsilon^2$ it can be written as

$$W(x) = \frac{1}{2R\varepsilon} \left(\frac{x^2}{1-x^2} + \mathrm{Ln} \frac{1}{1-x^2} \right) \tag{22}$$

It is not difficult to calculate the interaction energies between all the distribution of charges (anions and cations); these calculations lead to an expression for the electrostatic energy G_e per charge.

$$G_e = \frac{e^2}{2R\varepsilon} \left(-1 + \int_0^1 dx\, \Gamma(x) + (N-1) \int_0^1 dx \frac{\Pi^2(x)}{x^2} \right) \tag{23}$$

$\Pi(x)$ is the integral of $\Gamma(x)$ $\qquad \Pi(0) = 0 \quad \Pi(1) = 1$

$$\Pi(x) = \int_0^x dx\, \Gamma(x) \tag{24}$$

For Equation 23 the anions are considered to be "painted" on the sphere. This means that the attraction energy of one anion by the vesicle, which for $x = 1$ is infinite for a mathematical point charge, has been replaced by a finite energy of absorption. This energy is important, probably on the order of several tens of kcal, but as we shall suppose that all the anions are in contact with water, it turns out that the corresponding term is constant for a given sample and does not play any role in the total free energy. The entropy of the cations is computed from the classical formula (16):

$$S = - k_B \sum_i Pi \ln Pi \tag{25}$$

where the sum is extented over all states i with probability p_i. One gets, per cation:

$$G_s = - TS = k_B T \left(\int_0^1 dx\, \Gamma(x) \ln \frac{\Gamma(x)}{x^2} - 3 \ln R + \mathrm{Ln}\, N - 1 \right) \tag{26}$$

For given R and N, the distribution $\Gamma(x)$ is obtained by minimizing the functional

$$F = \frac{e^2}{2R\varepsilon} \left(T_1 + T_2 + T_3 \right) \tag{27}$$

$$T_1 = - 1 + \int_0^1 dx\, \Gamma(x)\, W(x) \tag{28}$$

$$T_2 = (N-1) \int_0^1 dx \frac{\Pi^2(x)}{x^2} \tag{29}$$

$$T_3 = B \int_0^1 dx\, \Gamma(x) \ln \left(\frac{\Gamma(x)}{x^2}\right) \qquad (30)$$

$$\frac{2k_B TR\varepsilon}{e^2} = B \qquad (31)$$

Equation 27 is non-linear. By variation calculus, its solution may be shown to satisfy the differential equation:

$$-\frac{dW}{dx} + 2\,(N-1)\,\frac{\Pi(x)}{x^2} - B\,\frac{d}{dx}\,\text{Ln}\left(\frac{\Gamma(x)}{x^2}\right) = 0 \qquad (32)$$

Although Equation 32 cannot be solved analytically, it can be the basis of numerical iteration, where a "guess" for $\Pi(x)$ is first introduced, from which $\Gamma(x)$ is deduced, this gives a second approximation for $\Pi(x)$, etc. For a given B, the procedure converges rapidly for a range of N which is quite enough for the vesicles we have to deal with.

R and N are related to ν if the volume of the cations is neglected, v_o is the volume of one water molecule:

$$\frac{4}{3}\,\Pi\, R^3 = N\, \nu\, v_o \qquad (33)$$

so that $B = \dfrac{k_B T}{e^2/2\varepsilon}\left(\dfrac{3}{4\eta}\, v_o\right)^{1/3} N^{1/3}\, \nu^{1/3}$ (34)

In Figure 4 we have plotted $\gamma(x)$ for R = 20 Å, T = 300 K, $\sigma = 20\ (\text{Å})^2$ and several values of N. The building of a layer of cations, at the surface of the vesicle, is clearly shown. In Figure 5 is plotted G, for the numerical values quoted above and several values of ν:

$$G = F + B\ (-\,3 \ln R + \ln N - 1) \qquad (35)$$

Two important conclusions may be drawn. First at a given N, G decreases if ν increases, showing a tendency for the vesicle to get more hydrated. However, for a given ν, G increases if N increases. This means that if we consider only the electrostatic and entropic contribution, an isolated sample (with ν constant) should split into small vesicles in order to minimize its free energy.

Interfacial Energy and Compact Coating. Until now we have not considered the interfacial energy of the vesicle and the matrix. If N is the number of anions coating the vesicle, the surface they occupy is $2\,N\,\sigma$, where σ is, as in the first part of this chapter, the cross-section of one segment of the polymeric chain. The remaining surface has to be coated with neutral monomers. Let us call γ_1 and γ_2 the specific interfacial energy of water with neutral material and charged anions, respectively. The total interfacial energy of one vesicle can be written as

$$2\,\sigma\, N\, \gamma_2 + (4\pi R^2 - 2\sigma N)\gamma_1 = 2\,\sigma\, N\,(\gamma_2 - \gamma_1) + 4\pi R^2 \gamma_1 \qquad (36)$$

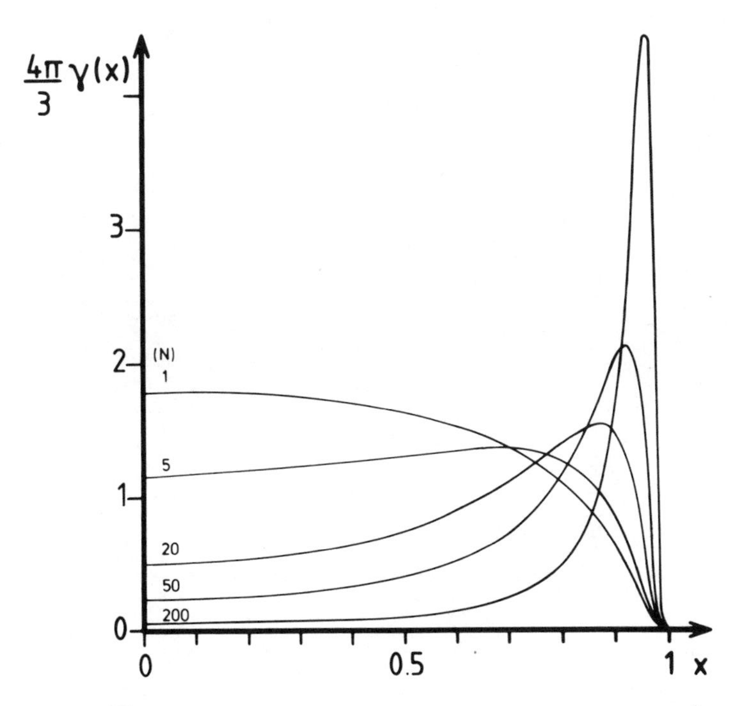

Figure 4. $\frac{4\Pi}{3}\gamma(x)$ plotted as a function of x for R = 20 Å and for several values of N. 1 would correspond to a uniform density for the cation. The progressive building of the bilayer is evident. (Reproduced from Ref. 5. Copyright 1985, American Chemical Society.)

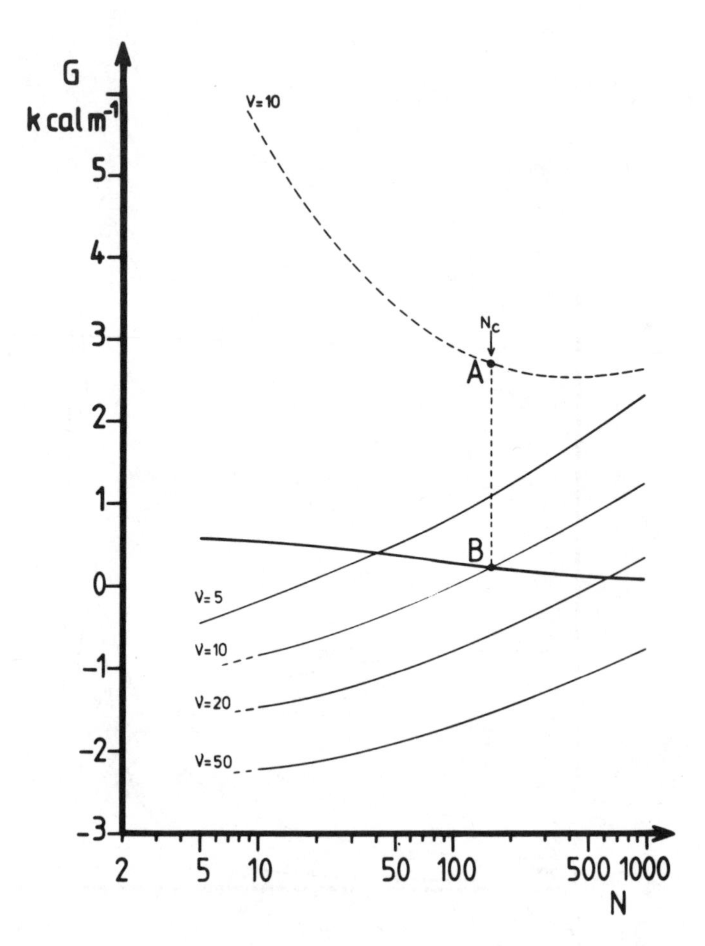

Figure 5. —— G from Equation 35 plotted as a function of N for different values of the hydration parameter γ.

---- The contribution of the interfacial energy (Equation 38) has been added to G for $\nu = 10$. One sees the minimum at $N \sim 500$. Point A, at the left of this minimum, represents the compact coating of the vesicle (Equation 39). AB represents the constant interfacial energy.

═══ G_c, free energy for the compact coating. A constant equal to AB has been neglected because the interfacial energy is constant in this case.

Per cation, this leads to a quantity that must be added to Equation 35:

$$2\,\sigma(\gamma_2 - \gamma_1) + 4\Pi\,\frac{R^2}{N}\,\gamma_1 \qquad (37)$$

γ_2 is related to the energy of absorption of the anions; it is large and negative. However, as it has already been said, it is a constant since we have supposed that all the anions are in contact with water. Therefore the first term in Equation 37 does not play any role. The second term is positive. A typical value for polyethylene (17) is $\gamma_1 \sim 40$ cgs. This relatively high value expresses the hydrophobic character of the contact of this material with water. Using Equation 37, this term, per cation, is

$$4\pi\left(\frac{3}{4\pi}\,v_o\right)^{2/3}\gamma_1\,\frac{\nu^{2/3}}{N^{1/3}} \qquad (38)$$

and it clearly favors large N.

We have added this contribution to G in Figure 5 (dashed line), for $\nu = 10$. For an isolated sample (ν = constant) the total free energy (Equation 35) plus (Equation 38) decreases now when N increases until it reaches a minimum where the effect of G (electrostatics and entropy) again predominates and the total free energy starts to increase with N. Can we conclude that, at thermal equilibrium, the vesicle's number of charges is fixed by the above minimum ? One has to pay attention to the fact that, because of geometrical hindrances, the number of anions cannot exceed a limiting value N_c, which for a given R is

$$4\,\Pi\,R^2 = 2\,N_c\,\sigma \qquad (39)$$

If we use the values for polytethylene, we see that N_c is smaller than the value of N corresponding to the mathematical minimum of the total free energy. This fact is still true for other values of ν, as long as γ_1 is large enough. The conclusion is straightforward: the hydrophobicity of neutral monomers causes the vesicle to be coated with the maximum N_c of anions, compatible with their cross-section.

It is quite evident at this stage, that although the interfacial term plays the fundamental role, it gives a constant contribution to the free energy per cation and may be dropped in thermodynamical calculations (with the restriction that γ_1 could depend, for instance, on T and give rise to a contribution to the specific heat).

We have plotted in Figure 5 the function G_c of N corresponding to this compact coating situation. It must now be recalled that the two relations, Equations 33 and 39 are simultaneously valid or that

$$N_c = 2\pi\left(\frac{3}{2}\right)^2\frac{v_o}{\sigma^3}\,\nu^2 = 1.54\,\nu^2 \qquad (40)$$

$$R = \frac{3}{2}\,\frac{v_o}{\sigma}\,\nu = 2.25\,\nu \qquad (R \text{ in } \mathring{A})$$

We can conclude that the size and capacity of the vesicle increase with the degree of hydration ν.

Hydration of Ionomers The knowledge of G_c is of course fundamental for the understanding of the thermodynamics of hydration of ionomers. We can calculate from it the chemical potential μ of water by deriving $\frac{dGc}{d\nu}$; we do that from an approximation of G_c

$$G_c = -0{,}71 + \frac{4.7}{N^{1/2}} - \frac{10}{N} = -0.71 + \frac{3.79}{\nu} - \frac{6.49}{\nu^2} \quad (41)$$

$$\text{or,} \quad \Delta\mu = -\frac{3.79}{\nu^2} + \frac{12.98}{\nu^3} \quad (\text{kcal m}^{-1}) \quad (42)$$

This function has a negative minimum for $\nu \sim 5.14$ and stays negative when ν increases.

The classical formula for a liquid containing a concentration $^1/\nu$ of solute inert molecules would give

$$\Delta\mu = -k_BTc = -\frac{k_BT}{\nu} = -\frac{0.6}{\nu} \quad (\text{kcal})$$

In comparison with Equation 42, we can see the error of using, without precautions, the formula for an ideal solution.

In the absence of any matrix contribution, one can get the law for the degree of hydration. The sample is in contact with a vapor at pressure p (18); P_c is the saturated vapor pressure.

$$kT\ \text{Ln}(P/P_c) = \Delta\mu \ \text{or,}\ P = P_c \exp\left(-\frac{6.36}{\nu^2} + \frac{21.78}{\nu^3}\right) \quad (43)$$

Of course, any deviation from this law should reflect the influence of the matrix. Within the limited scope of this paper, we shall not enter into this problem. Qualitatively, what can be said here is that, following the increase of N_c, and of D (the intervesicle distance), the tension of the "connecting" segments increases which results in an entropic term that adds to G_c. At small c, where the vesicles are well separated, this new entropic term can stop the process of hydration at a finite value of ν. At large c, the vesicles are almost in contact, even in the dry state. The process of hydration accentuates this effect and the vesicles build a network of channels. In this situation, the assumption of sphericity for the vesicles is, of course, no longer valid.

Acknowledgments

It is a pleasure to thank Dr M. Pineri and F. Volino from Grenoble and Pr A. Eisenberg from McGill University, Montreal, for fruitful discussions.

References

(1) Holliday, L. Ed."Ionic Polymers" ; Halstead Press, New York (1977).
(2) Eisenberg A., King M., "Ion Containing Polymers" ; Academic Press, New York (1978)
(3) Macknight W.J., Earnest T.R., Jr ; Polym. Sci. Macromol. Rev. (1981) 16, 41

(4) Hopfinger A.J., Mauritz K.H., "Comprehensive Treatise of Electrochemistry" Vol. 2, chap. 9 p. 521-535, J. O'M. Brockris, B.E. Conway, E. Yeager and R.E. White, Plenum, New York, (1981)
(5) Dreyfus B.; Macromolecules (1985) 18, 284
(6) Eisenberg A.; Macromolecules (1970) 3, 147
(7) Stratton J.A., Electromagnetic Theory, Mc Graw-Hill, New York (1941) p. 176
(8) Marx B.C.M., Caulfield D.F., Cooper S.L. ; Macromolecules (1973) 6, 344
(9) MacKnight W.J., Taggart W.P., Stein R.S. ; J. Polym. Sci. Polym. Symp. (1974) 45, 113.
(10) Wilson F.C., Longworth R., Vaughan D.J., Am. Chem. Soc. Polym. Prepr. (1968) 9, 595
(11) Yarusso D.J., Cooper S.L. ; Polym. Prepr. (1981)
(13) Gierke T.D., Hou W.Y., Perfluorinated Ionomers Membranes ; Eisenberg A., Yeager H.L., Ed. ; ACS Symposium Series 180 ACS, Washington (1982) p. 283
(14) Dreyfus B., to be published
(15) Landau and Lifschitz, Electrodynamics, Pergamon Press, New York (1960), p. 17
(16) Kittel C. ; Elementary Statistical Physics, Wiley, New York (1958) p. 5
(17) Petke F.D. and Ray B.R. ; Coll. Interface Sci.(1969) 31 p. 316
(18) Dreyfus B., 1983, Polym. Sci. Phys. Ed. (1983) 21, 2327

RECEIVED September 25, 1985

9

Composite Nature of Ionomers

Properties and Theories

William Y. Hsu

Central Research and Development Department, E I du Pont de Nemours and Company, Experimental Station, Wilmington, DE 19898

The composite nature and its influence on physical properties of ionomers and relevant theories are reviewed. Using perflourinated ionomers as examples, ion clustering and its energetics, percolation phenomena in transport and elastic properties, and morphological effects on ion selectivity are examined.

Ionomers have been studied extensively during the past twenty years (1). An important conclusion is that these materials are generally heterogeneous. The ionic groups segregate into multiplets, aggregates, clusters or inverted micelles instead of being uniformly distributed (2). On the one hand, ion clusters in dry samples are driven electrostatically since the total energy is lowered substantially by the formation of electric multiplets (3). On the other hand, inverted micelles are formed in hydrated samples due to the competition between hydrophilic acid side groups and hydrophobic polymer backbones (2). This paper focuses on the multiphase, or composite, nature of ionomers and its influence on physical properties.

To be specific, perfluorinated ionomers of the form

$$[(CF_2)_n\underset{|}{C}F]_m \quad OR_f\text{-}SO_3X \text{ (or } CO_2X\text{, etc.)}$$

will be used exclusively, but the concepts discussed here are very general and applicable to other ionomers. In the above formula, n is 6 to 13, R_f is a perfluoro alkylene group that may contain ether oxygen and X is any monovalent cation. The perfluoro side chain may contain sulfonate, carboxylate or sulfonamide group and the ionomer will be designated accordingly. The number of ionic side groups is characterized by the equivalent weight (EW) which is defined as the dry mass in grams of the acid form of the ionomer that is needed to neutralize one equivalent of base. The perfluorinated ionomers are known for their electrochemical properties and various applications, and have been studied extensively (4).

The two phase nature of perfluorinated ionomers was first observed by Gierke and his collaborators using small angle x-ray

0097-6156/86/0302-0120$06.00/0

scattering (SAXS) (5,6) and electron microscopy (6,7). They reported that scattering from the sulfonyl fluoride precursor polymer was very weak and without any noticeable features. However, a new SAXS peak representing ion clustering appears in dry sulfonic acid ionomers. Similar peaks had previously been observed in the dry state of other ionomers (8). In hydrated perfluorinated ionomers, this SAXS peak grew in intensity and shifted to smaller scattering angle. It is interpretted as a Bragg-like diffraction peak and corresponds to a center-to-center separation of about 50A between clusters for a 1200EW sulfonate ionomer. These SAXS features were subsequently confirmed by Roche *et al* (9) although details of their interpretation were somewhat different. They also showed that the electron density fluctuations deduced from SAXS data were only consistent with a complete separation of the aqueous phase from the polymer. The existence of ion clusters was further supported by electron micrographs obtained from dry sulfonate ionomers stained by heavy counter ions (6,7). In the following, I shall outline an elastic theory for cluster formation in hydrated samples (10), discuss the percolation theory of ion transport (11,12), apply the effective medium theory (13) to model elastic properties (14) and highlight the influence of morphology on ion selectivity (15,16).

Elastic Theory for Ion Clustering

Qualitatively, the equilibrium cluster diameter d_c is determined by the balance of three terms: (i) elastic deformation of the medium, (ii) difference in the condensation energies of the aqueous phase before and after entering the polymer and (iii) hydrophilicity of the ion-exchange groups. The first two terms are bulk contributions and proportional to the volume of the ion cluster; the last term originates from the water-polymer interface where the ion-exchange groups reside and is proportional to the surface area of the cluster. Energetically, the first term is a barrier that needs to be overcome whereas the last term is the primary driving force of cluster formation in hydrated samples. The elastic contribution can be computed from known tensile modulus but the other two terms are best determined empirically (10). In this regard the theory is semi-phenomenological and differs from a previous molecular approach (2).

The starting point of the theory is a hypothetical dry cluster of diameter d_o (typically 2 nm) and N_p ion exchange sites embedded in an otherwise hydrated sample. The objective is to follow the change in free energy associated with the growth of this cluster while holding N_p constant. In the dry state, the following contributions to the free energy are important: the elastic energy needed to exclude the polymer from the interior of ion cluster, ion-ion and ion-polymer interactions, and the water-water interaction outside the ionomer. The last term is included to allow for changes in vapor pressure and chemical potential of the external water source. As the dry cluster hydrates, various water related contributions must be added to the free energy. It is convenient to separate the cluster into three regions: the first hydration shell, the second hydration shell and the bulk interior of the cluster as shown in Figure 1. Water molecules in the first hydration shell interact with the fluorocarbon matrix, the ion exchange groups and other water molecules; but water molecules in the interior of the cluster interact with each others only. The net change is then minimized to obtain the equilibrium cluster diameter d_c (10):

$$d_c = \frac{4\sigma_w K_s}{[2/3]E(f)[1-d_o^2/d_c^2]+K_1} - 2\sigma_w' \quad (1)$$

where E(f) is the tensile modulus and f is the aqueous content of the hydrated ionomer, σ_w is 0.31 nm and represents the diameter of a water molecule, σ_w' is a normalized diameter that is slightly less than σ_w; the slight difference between these two diameters will be ignored in actual computation. The experimental E(f) for a 1200EW sulfonate ionomer (17) is represented by the dotted line of Figure 2 and the significance of the theoretical curve will be discussed later. In Equation 1, K_1 and $\sigma_w K_s$ represent the condensation and surface cotnributions, respectively. The reference state for $-K_1$ is the bulk condensation energy of free water and, therefore, K_1 depends on the chemical potential of the external water source. To a first approximation, K_1 is independent of the counter cations, K_s is independent of the surface concentration of an ionic exchange sites but increases with their hydrophilicity, all of which are consistent with data (10). Finally, the effect of counter cations is implicitly contained in the tensile modulus since it depends on the water content which, in turn, is affected by the cation form (17). When applied to sulfonate ionomers, Equation 1 faithfully describes the observed variation in cluster diameter with cation form, equivalent weight and water content as shown in Figure 3. Notice the pronounced effect of chemical potential in the water content dependence. From a least square fit to the EW and cation form data we found K_s=917 joule-cm^{-3} and K_1=170 joule-cm^{-3}, which may be compared to the heat of water vaporization 2260 joule-cm^{-3}, the thermal energy 138 joule-cm^{-3} per water molecule at 300 K, and the formation energy of a hydrogen bond 418 joule-cm^{-3}. The free energy of "bulk" water inside a hydrated ionomer is slightly larger than pure water because K_1 is small and positive (since the reference state is with respect to $-K_1$). But this effect is more than offset by the much larger lowering in free energy produced by the surface interaction K_s. Accordingly, the equilibrium cluster diameter is basically determined by the balance between the hydrophilic surface interaction and the elastic deformation of the fluorocarbon matrix.

The thermodynamic stability of the channels connecting adjacent ion clusters has also been analyzed (10). These channels were first proposed by Gierke to explain his transport data (5,17). Using the same elastic theory as discussed above, the equilibrium channel diameter was found to be (1.4 ± 0.2) nm in excellent agreement with experimental values of 1.2 to 1.3 nm. The change in free energy associated with channel formation was found to be −11 joules-cm^{-3}. Thus these channels are stable but are forming and unforming continually at ambient temperature.

Percolation Theory for Ion Transport

For transport considerations the most crucial factor is the formation and random distribution of conductive ionic clusters in a

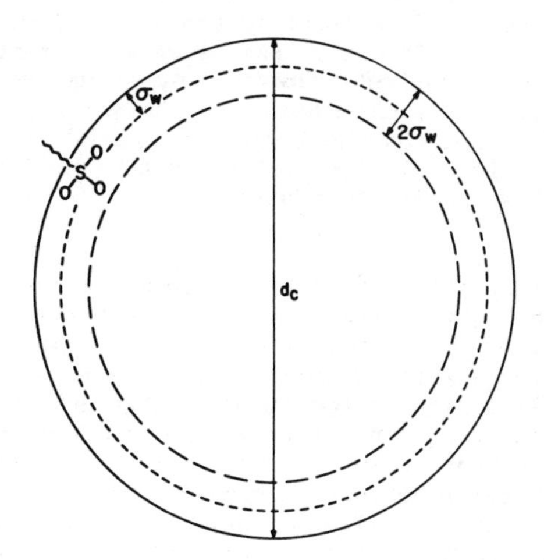

Fig. 1 Schematic of a hydrated ion-cluster. The three regions indicated are: (i) the first hydration shell, (ii) the second hydration shell and (iii) the cluster interior, respectively. Reproduced with permission from Ref. 10, Fig. 1. Copyright 1982, American Chemical Society.

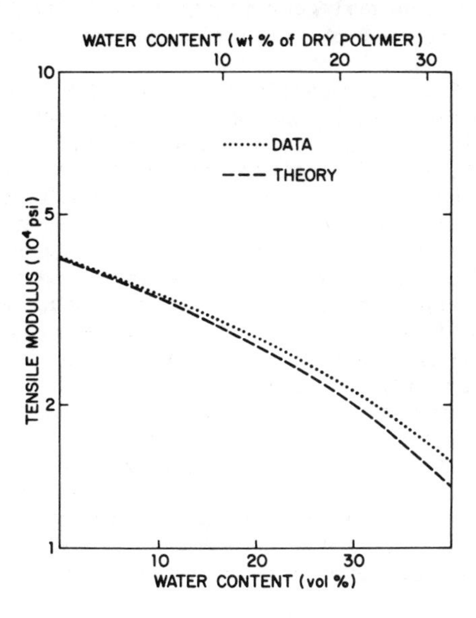

Fig. 2 Tensile modulus of 1200 EW perfluorosulfonate ionomer. Reproduced with permission from Ref. 16, Fig. 2. Copyright 1984, Electrochemical Society.

fluorocarbon matrix (7,12). At low water content, the clusters are completely disjointed and macroscopic transport of sodium and hydroxyl ions is very difficult. As the water content increases, the aqueous clusters progressively connect into an extended network that eventually pervades the whole sample. A threshold f_o exists between these limits and demarcates the insulator-to-conductor transition. The conductivity σ is negligible below the threshold and increases steadily above it according to a power law (11):

$$\sigma=\sigma_o(f-f_o)^t \tag{2}$$

where t is an universal constant that depends on the spatial dimensionality and σ_o is a prefactor. The best theoretical value for t is about 1.7 for a three dimensional system. Empirically, f_o depends not only on spatial dimensionality but also on mixing conditions, particle size, shape and distribution, and the prefactor σ_o depends on details of ionic interactions and diffusion processes as well. The important feature is that all topological and geometrical information has been lumped together in the $(f-f_o)^t$ factor. Therefore the complicated many-cluster transport problem has been reduced to a consideration of the prefactor σ_o which is controlled by ion-ion and ion-water interactions within a cluster and diffusion processes between two adjacent clusters only.

It is interesting to note that the shear modulus of hydrated samples will decrease during hydration since the aqueous phase can not and does not support any shear stress. However, changes in elastic moduli remain moderate and gradual at f_o since the continuous polymer phase still provides the necessary shear rigidity. The percolative transition analogous to Equation 2 for tensile and shear moduli occurs at such a high aqueous content that the continuous polymer phase has begun to break up.

The applicability of Equation 2 to perfluorinated ionomers has been tested experimentally. One of the methods used to measure σ was designed by Berzins (19) and is shown schematically in Figure 4. Since the conductivity of electrolytes and the cross section and thickness of the membrane are known, σ can be determined from the voltage drops across the three pairs of probe electrodes 1-2, 3-4 and 5-6. The sodium current efficiency (CE) can also be determined by titrating the amount of caustic soda generated over a given period of time. The confinement chambers around the working electrodes are used to eliminate free bubbles near the membrane. Our normalized transport data for sulfonate, carboxylate and sulfonamide ionomers are plotted in Figure 5; the universal percolative nature of perfluorinated ionomers can be clearly seen. The prefactor σ_o is $0.04\Omega^{-1}cm^{-1}$ for sulfonamide and $0.36\Omega^{-1}cm^{-1}$ for sulfonate ionomers. The exponent t is 1.5 ± 0.1 in reasonable agreement with theory and the thresholds are between 8 to 10 vol. %, which are consistent with the bimodal distribution in cluster size postulated by the cluster-network model (5,18). This theory has also been applied recently to delineate sodium selectivity of perfluorinated ionomers (20).

Effective Medium Theory for Elastic Properties

The normal range of aqueous content in a wet ionomer is sufficiently far from the elastic percolative transition that an effective medium

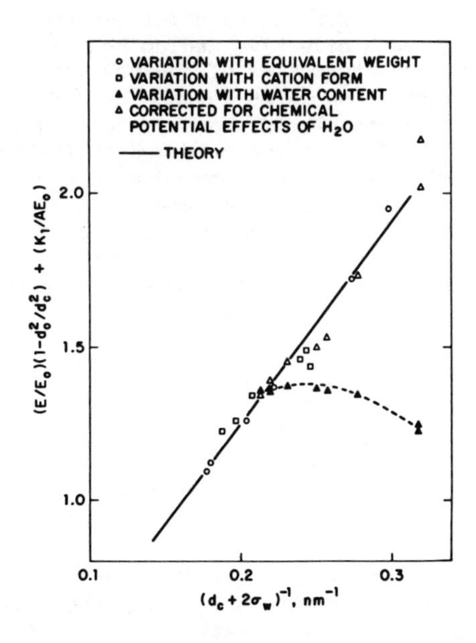

Fig. 3 Variation of equilibrium cluster diameter d_c with EW, cation form and water content, where E_o=275 joule-cm^{-3} is the tensile modulus of a dry, 1200 EW sulfonate ionomer, A=0.667 is a constant and d_c is obtained from SAXS and water sorption data. The solid line is a least square fit of Eq. 1 to the EW and cation form data.

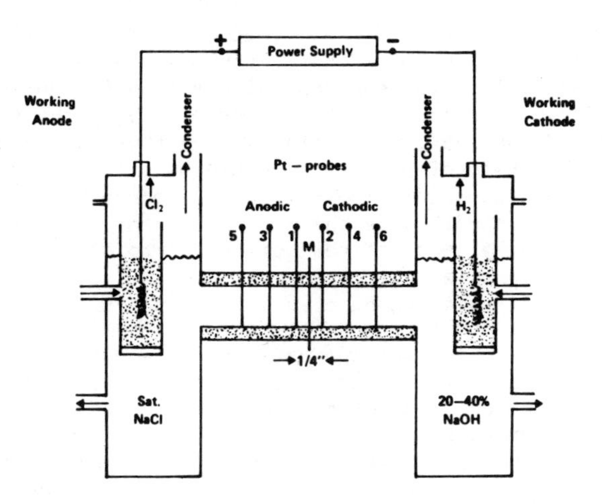

Fig. 4 Electrochemical scheme of measuring the ionic conductivity of perfluorinated ionomer membranes.

theory (13) is adequate. This theory simulates properties of a composite by a homogeneous effective medium that is determined self-consistently by the properties of the original components. In the present context, the task is to relate the effective shear and bulk moduli, G and K, respectively, to the actual moduli of the two phases, G_j's and K_j's. This is accomplished as follows. Changes in the stress and strain tensors are computed when a small spherical portion of the effective medium is replaced by one of the component materials of the composite. The average changes in these quantitites are then required to vanish since the effective medium supposedly has identical properties as the original composite. We thus obtain the following self-consistent equations (14):

$$\sum_{j=1}^{2} f_j \left(\frac{K_j}{K}\right)\left(\frac{3K+4G}{3K_j+4G}\right)=1 \qquad (3a)$$

$$\sum_{j=1}^{2} f_j \left[\frac{G-G_j}{(7-5\nu)G+(8-10\nu)G_j}\right]=0 \qquad (3b)$$

where f_j is the volume fraction of the jth component and ν is the Poisson ratio of the effective medium. Using the known relations between elastic moduli and known elastic constants for water (21) and dry perfluorosulfonate ionomer (17), the tensile modulus E of hydrated samples has been computed (16). It is compared to experimental data (17) in Figure 2 and the agreement is good.

Morphological Effects in Ionomer Blends

Different perfluorinated ionomers have very different transport properties and Na selectivity. For example, the Na^+ and OH^- conductivities are $1.80\times10^{-2}\ \Omega^{-1}\ cm^{-1}$ and $1.36\times10^{-2}\ \Omega^{-1}\ cm^{-1}$, $6.99\times10^{-3}\ \Omega^{-1}\ cm^{-1}$ and $6.10\times10^{-4}\ \Omega^{-1}\ cm^{-1}$, and $1.44\times10^{-3}\ \Omega^{-1}\ cm^{-1}$ and $2.50\times10^{-4}\ \Omega^{-1}\ cm^{-1}$ for 1100EW sulfonate, carboxylate and sulfonamide ionomers, respectively. The corresponding sodium CE, which is the fractional contribution to the total conductivity by the Na^+ ions alone, is 57%, 92% and 85%, respectively. We thus wonder whether interesting combination of properties could be attained in blends (15,16). In this context, morphology is an important parameter (15) because it can be controlled by the viscosities of the component phases. Theoretically, most of the common morphologies can be simulated with ellipsoids having properly chosen semi-principal axes a, b and c. An ellipsoid degenerates into a spheroid when two of the three axes are equal and into a sphere when all three axes are equal. Usually spheroids are adequate approximations because oblate spheroids (a=b>c) naturally cover spherical (a=c) to lamellar (a>>c) morphologies and prolate spheroids (a>b=c) cover spherical (a=c) to fibrillar (a>>c) morphologies. These spheroids are oriented in such ways that the current flows either along the short axis c for the oblate case or along the long axis a for the prolate case. Empirically a and c are several μm and much larger than the equilibrium cluster diameter d_c (∿5nm). Thus each spheroidal domain contains several million ion-clusters and behaves just like any typical bulk sample.

Transport properties of ionomer blends, characterized by a given type of spheroids and the aspect ratio, c/a, can now be analyzed by the effective medium theory discussed in the previous section. In this theory, the two phases are assumed randomly mixed and the probability of finding each phase is equal to its volume fraction f_i. The effective conductivity, σ, of the composite for either Na^+ of OH^- ions is given by (15):

$$(1-F)\sigma^2+[\sigma_1(F-f_1)+\sigma_2(F-f_2)]\sigma-\sigma_1\sigma_2F=0 \qquad (4)$$

where σ_i is the appropriate Na^+ or OH^- conductivity for phase i (=1 or 2). Morphological information enters Equation 4 via the factor F. It is exactly 1/3 for spheres, 1 at the lamellar limit and 0 at the tubular limit as shown in Figure 6. Notice that Equation 4 is symmetric with respect to the two phases and thus it will automatically reverse their roles (majority vs. minority) at the medium composition.

The computed CE as a function of blend composition is plotted in Figure 7 for various morphologies. The dashed curve is the spherical limit. The CE increases very slowly and >80 vol. % of carboxylate is needed to approach the selectivity of pure carboxylate. Physically, this slow turn around can be understood as follows. Since the OH^- conductivity is 22 times higher in the sulfonate phase than the carboxylate phase and since the penalty of going around a sphere via its circumference instead of diametrically across is only π, the OH^- ions can easily bypass the highly selective carboxylate domain via the surrounding sulfonate phase. These shunt paths are cut off only at high carboxylate contents in the spherical limit. The situation is even worse with oriented prolate spheroids because the longitudinal circumference approaches the length of the long principal axis, 2a, rapidly as the spheroids elongate. The geometric penalty of going around the carboxylate phase is thus even less, leading to much poorer CE. In contrast, the longitudinal circumference of an oblate spheroid increases rapidly with respect to the transverse short principal axis, 2c, as the spheroid flattens out. The OH^- ions soon find themselves taking such long tortuous pathways to bypass the carboxylate phase that the conductivity gain through the sulfonate phase is totally negated. The sodium CE thus improves dramatically as shown by the top two curves of Figure 7. Our experimental data normally fall inside the bar regions of Figure 7 suggesting strongly lamellar carboxylate domains. This was subsequently confirmed by electron microscopy (15). The predicted difference in CE between lamellar and spherical morphologies (86% vs 62%) has also been observed (89% vs 56%) in a 25% carboxylate blend of 1100EW ionomers (15).

We also wonder whether the CE of these blends can be improved as sulfonamide has much higher CE than sulfonate. Theoretically, however, the CE for lamellar sulfonamide/carboxylate blends has a totally different curvature (16) as shown by the dashed curve of Figure 8. The CE rises slowly and stays close to the behavior of pure sulfonamide for most of the composition. Beyond 25% carboxylate, the CE of a lamellar sulfonamide/carboxylate blend is

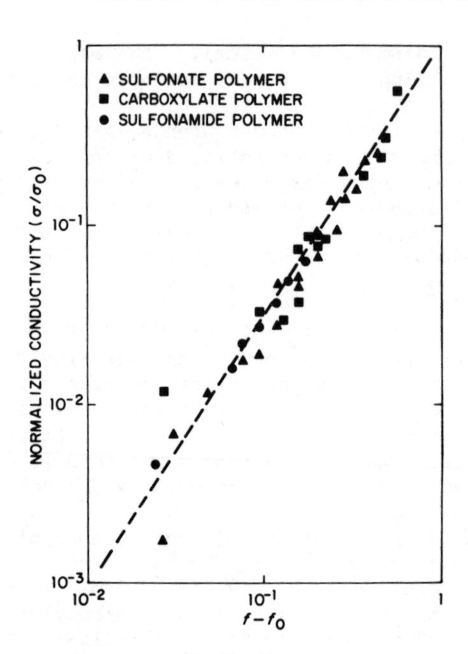

Fig. 5 Universal plot of normalized conductivity (σ/σ_o) vs excess aqueous content ($f-f_o$). The dashed line is a least square fit of Eq. 2 to data. Reproduced with permission from Ref. 16, Fig. 1. Copyright 1984, Electrochemical Society.

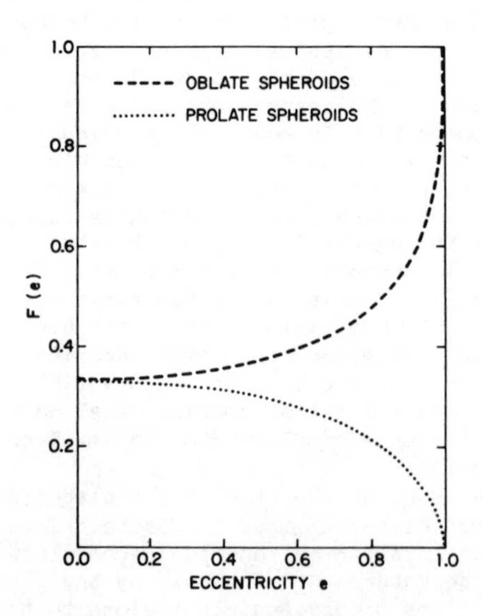

Fig. 6 Eccentricity dependence of the F function. The eccentricity e of a spheroid is related to the aspect ration (c/a) by: $e^2=1-(c/a)^2$. Reproduced with permission from Ref. 16, Fig. 4. Copyright 1984, Electrochemical Society.

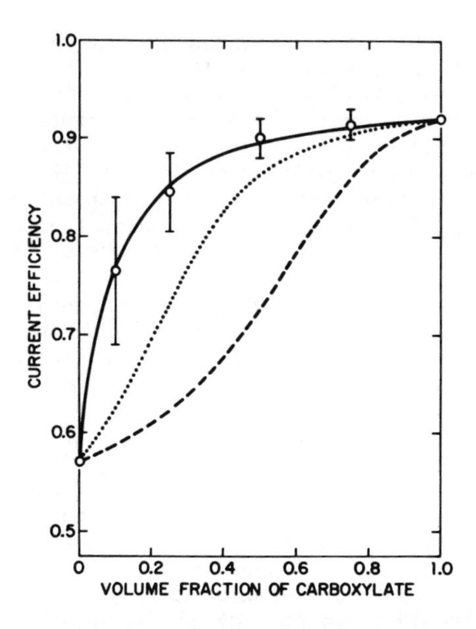

Fig. 7 Current efficiency of carboxylate/sulfonate ionomer blends. Curves are predicted behavior of oriented oblate spheroids whose aspect ratios are 0.01 (top curve), 0.25 (middle curve) and 0.995 (bottom curve), respectively. Reproduced with permission from Ref. 15, Fig. 2. Copyright 1983, American Chemical Society.

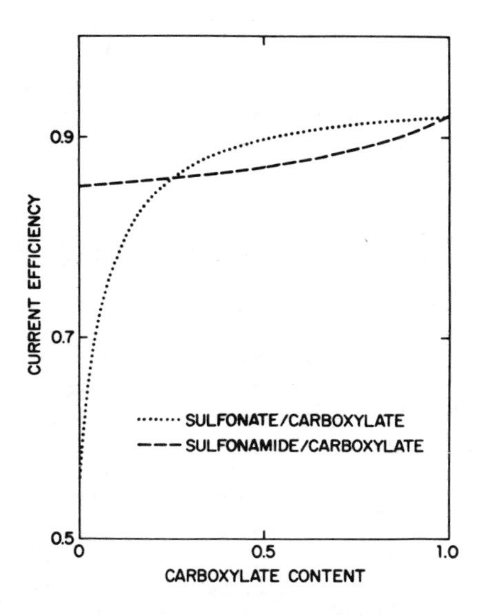

Fig. 8 Computed current efficiency for two different lamellar, carboxylate blends. The equivalent weight is 1100. Reproduced with permission from Ref. 16, Fig. 8. Copyright 1984, Electrochemical Society.

actually lower than that of a similar sulfonate/carboxylate blend. For example, at 50% composition, the CE is 86% (vs 87% theoretically) for the former and 91% (vs 90% theoretically) for the latter blend (16). Physically the unusual change in the curvature of the CE curves arises from differences in the transport characteristics and roles of the sulfonate and sulfonamide phases. In a sulfonate/carboxylate blend, the carboxylate phase is the selective component but the sulfonate phase is the conductive component. The lamellar morphology works because it forces the OH^- ions through the selective carboxylate domain which would otherwise be bypassed. In a sulfonamide/carboxylate blend, however, the carboxylate phase is both the selective and the conductive component. The lamellar morphology consequently hurts the performance by diverting the OH^- ions across the sulfonamide phase unnecessarily.

Summary

In summary, I have discussed a semi-phenomenological elastic theory for ion clustering in ionomers. The theory is consistent with observed trends in perfluorinated ionomers. I have also demonstrated the percolative nature of ion transport in these ionomers and computed quantitatively their tensile modulus. Finally, I have discussed the influence of morphology on ion selectivity in perfluorinated ionomer blends. In particular, I have pointed out that an universally preferred morphology beneficial to all blends does not exist; the ideal morphology must be individually determined based on component properties. Most of the theories and conclusions here are very general and applicable to other composite polymer systems.

Acknowledgments

It is my pleasure to thank Drs. J. R. Barkley, T. Berzins, T. D. Gierke, P. Meakin, C. J. Molnar, and G. E. Munn for their many contributions during the course of this work.

Literature Cited

1. Bazuin, C. G.; Eisenberg, A. Ind. Eng. Chem. Prod. Res. Dev. 1981, 20, 271-86.
2. Mauritz, K. A.; Hopfinger, A. J. In "Modern Aspects of Electrochemistry"; Bockris, J. O'M.; Conway, B. E.; White, R. E., Ed.; Plenum: New York, 1982; No. 14, Ch. 6, pp. 425-508.
3. Eisenberg, A. Macromolecules 1970, 3, 147-54.
4. See, for example, "Perfluorinated Ionomer Membranes"; Eisenberg, A.; Yeager, H. L., Ed.; ACS SYMPOSIUM SERIES No. 180, American Chemical Society: Washington, D.C., 1982.
5. Gierke, T. D. J. Electrochem. Soc. 1977, 124, 319C.
6. Gierke, T. D.; Munn, G. E.; Wilson, F. C. J. Polym. Sci Polym Phys. Ed. 1981, 19, 1687-704.
7. Hsu, W. Y.; Gierke, T. D. J. Membrane Sci. 1983, 13, 307-26.
8. MacKnight, W. J.; Earnest, T. R., Jr. J. Polym. Sci. Macromol. Rev. 1981, 16, 41-122.

9. Roche, E. J.; Pineri, M.; Duplessix, R.; Levelut, A. M. J. Polym. Sci. Polym. Phys. Ed. 1981, 19, 1-11.
10. Hsu, W. Y.; Gierke, T. D. Macromolecules 1982, 15, 101-5.
11. See, for example, Zallen, R. "The Physics of Amorphous Solids"; John Wiley & Sons: New York, 1983; Ch. 4, pp. 135-204 and references therein.
12. Hsu, W. Y.; Barkley, J. R.; Meakin, P. Macromolecules 1980, 13, 198-200.
13. Wood, D. M.; Ashcroft, N. W. Philos. Mag. 1977, 35, 269-280 and references therein.
14. Hsu, W. Y.; Giri, M.; Ikeda, R. M. Macromolecules 1982, 15, 1210-2 and references therein.
15. Hsu, W. Y.; Gierke, T. D.; Molnar, C. J. Macromolecules 1983, 16, 1945-7.
16. Hsu, W. Y. J. Electrochem. Soc. 1984, 131, 2054-8.
17. Grot, W. G. F.; Munn, G. E.; Walmsley, P. N. J. Electrochem. Soc. 1972, 119, 108C.
18. Gierke , T. D.; Hsu, W. Y. In Reference 4, Ch. 13, pp. 283-307.
19. Berzins, T. J. Electrochem. Soc. 1977, 124, 318C.
20. Dayte, V. K.; Taylor P. L.; Hopfinger, A. J. Macromolecules 1984, 17, 1704-8.
21. "Lange's Handbook of Chemistry"; Dean, J. A., Ed.; McGraw-Hill: New York, 1978; Table 10-40, p. 10-122.

RECEIVED June 10, 1985

POLYMERIC MEMBRANES

10

Structure and Function of Membranes for Modern Chloralkali Cells

Ronald L. Dotson

Olin Corporation, New Haven, CT 06511

With the advent of dimensionally stable anodes and perfluorinated ion exchange membranes, over the past decade and a half, came an alternative method for the manufacture of chlorine and caustic soda. This new process produces a food grade product without pollution.

Perfluorinated membranes now provide us with the key to a new era of high technology in electrochemical science and technology, especially in the manufacture of heavy chemicals. These membranes can be characterized by their structure and function.

In the early 1970's a maximum in the cathode current efficiency was found to appear as a function of caustic strength, (1), using these perfluorinated membranes and this germinated the development of the first commercially successful chlor-alkali membrane cells. The maximum is thought to occur through discontinuous phase change zones less than ten microns thick on the cathode surface of the membrane. Here percolation can occur through topologically distorted clusters in high density films maintained under dynamic electrical load.

Two revolutionary developments have made a subtle, but permanent change in the technology of chlorine and caustic manufacture during the past decade and a half, dimensionally stable anodes and perfluorinated ion exchange membranes. New cell concepts are now made

0097–6156/86/0302–0134$06.00/0

possible using these components in unique geometric electrode-membrane designs which demonstrate lower operating costs than past technology without adding pollution problems associated with asbestos and mercury cell processes. The basic operational process for a simplified chlor-alkali membrane cell is illustrated in "Figure 1".

It is not at all surprising that the large producers of fluorine and fluorinated products have been in the vanguard of an intensive development effort to produce the most efficient, long-lived ion exchange membranes possible. Even though the complete history of these polymeric ion exchange membranes spans just over two and a half decades, some of the scientists and engineers in this field have become prolific on both the theory and practical applications. It is largely because of the large investment in time and manpower and the unique juncture of technology with policies and politics that the theories of ion exchange membranes are in a much more advanced stage than any of the other ion exchange systems.

Membranes can be characterized by their structure and function, that is how they form and how they perform. It is essential that the cation exchange membranes used in chlor-alkali cells have very good chemical stability and good structural properties. The combination of unusual ionic conductivity, high ionic selectivity and resistance to oxidative hydrolysis, make the perfluorinated ionomer materials prime candidates for chlor-alkali membrane cell separators.

Structure. The first commercially successful chlorine-caustic cells were developed and tested at Diamond Shamrock's T.R. Evans Research Center in Painesville, Ohio in the early 70's. The resin formulation for these separators was based on the polytetrafluoroethylene backbone with short polyether side chains as shown:

$$----(CF_2CF_2)-(OCF_2\underset{CF_3}{CF})_x-OCF_2CF_2SO_3H \qquad (1)$$

The equivalent weights for these resins ranged from 1000 to 2000 meq/g, in both sulfonic and carboxylic acid forms.

The semicrystalline, supermolecular structure of the organic carboxylate and the amorphous structure of the sulfonate resins have been studied with x-ray scattering and mechanical relaxation. This work shows no trace of crystallinity in the sulfonates, but the stress-relaxation data suggests the presence of a common structural feature, ion-clustered structure, with regions of high and low ion content. In "Figure 2" is shown the x-ray diffraction patterns depicting the supermolecular structure of perfluorocarboxylate and the sulfonate. Here is shown the amorphous halos in both

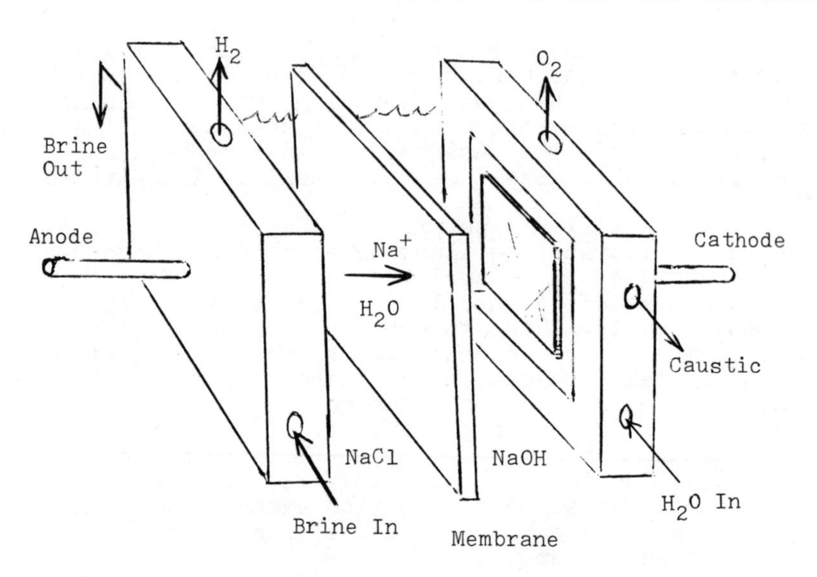

Figure 1. Basic operational process for membrane chloralkali cells.

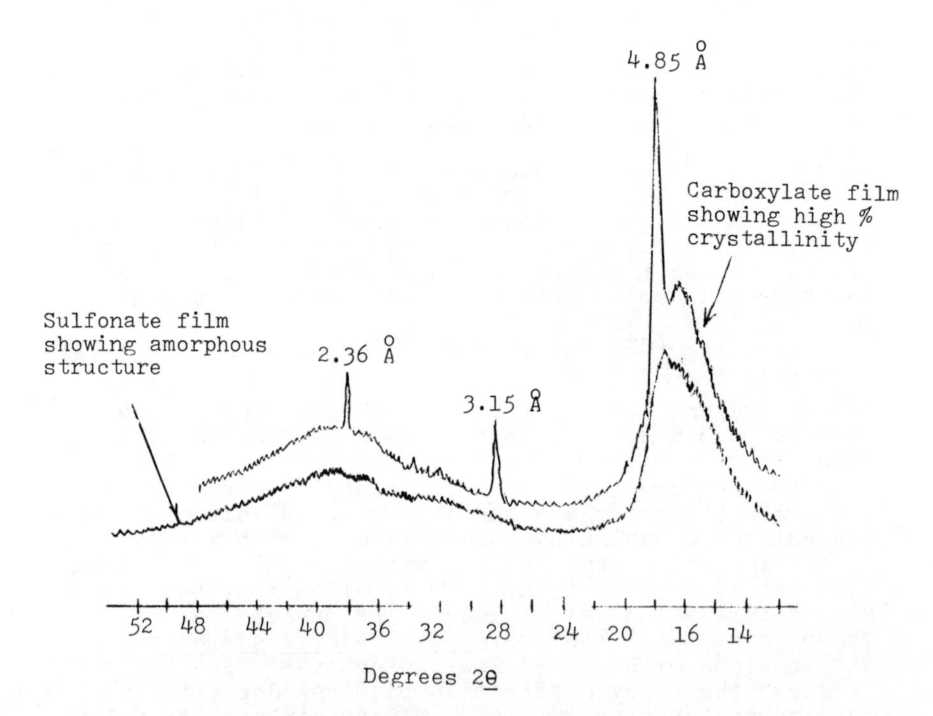

Figure 2. X-ray diffraction scans showing the supermolecular structures of perfluorocarboxylate and sulfonate materials.

of the polymers, and also the carboxyl group aggregation yielding crystalline peaks at 2.36, 3.15 and 4.85Å. In "Figure 3" we see an idealized model for a random-coil network of ion-clustered structure of an amorphous glass in a polymer chain. One polymer chain is darkened for better visualization. A concept of how ionic clusters form in the sulfonic acid polymer is presented here in "Figure 4". Here is shown the region of cluster sol and polymer gel proposed for the perfluorinated sulfonate membranes. The cluster region is the gate which controls ion flow through the separator, and it is modulated by concentration, temperature, structure and current density in these films.

In addition to the ion-clustered gel morphology and microcrystallinity, other structural features include: pore-size distribution, void type, compaction and hydrolysis resistance, capacity and charge density. The functional parameters of interest in this instance include permeability, diffusion coefficients, temperature-time, pressure, phase boundary solute concentrations, cell resistance, ionic fluxes, concentration profiles, membrane potentials, transference numbers, electroosmotic volume transfer and finally current efficiency.

When strong interactions take place between the membrane structure and solvated ions and solvent, the solubility of the permeant species are influenced by modification of the solvation capacity of the solvent molecules through a restricted binding caused by the close proximity of polymer substrate-pore walls and hydrated ions. In this instance the membrane forms a polymer-solvent complex in the thin controlling layer next to the catholyte which rejects hydroxyl ions. The brine side of this separator is a highly solvated gel. This effect becomes much more pronounced in regions where the alkali solution becomes most highly structured, as $NaOH \cdot 3\frac{1}{2}H_2O$ or $KOH \cdot 4H_2O$. In this case water associated with hydrophilic groups fills the flow channels between crystallites in the thin alkaline skin or layer film thus producing a lowering of the dielectric constant and thereby introducing anisotropic microporous characteristics there.

Dense, impermeable perfluorinated membranes are converted into permeable materials through increased swelling in a suitable medium. It is clear that the structure and function of membranes are interdependent, (2,3).

Function. The chemical, thermal and mechanical stability and ion-exchange behavior of the perfluorinated resins have been found to depend on the resin structure, degree of crosslinking and also on the nature and number of fixed ionic groups. The degree of crosslinking through ionic linkages in the clusters and covalent linking in the polymer backbone, established a mesh width for the

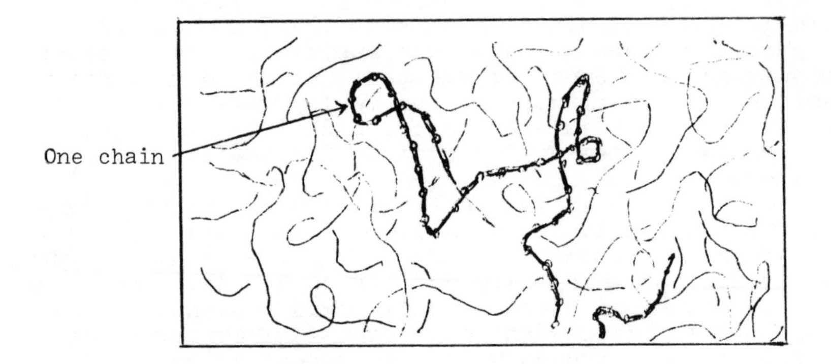

Figure 3. Idealized model of a random-coil network of ion-clustered structure for an organic glass.

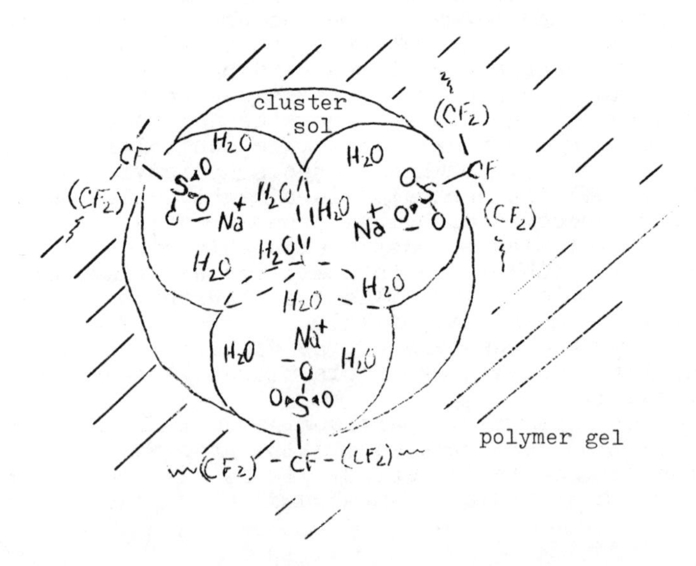

Figure 4. Ionic clusters formed in perfluorinated sulfonic acid membrane.

matrix and also swelling ability of resin and ionic mobilities for the counterions in the mesh. Many of the perfluorinated materials are straight chain polymers linking through ionic entanglement alone. The average mesh width of the highly crosslinked resins is of the order of only a few angstrom units. Shown in "Figure 5" is a conception of the network of interconnected channels in the macro gel-sol network. This depicts the approximate spacing between functional groups, where the mesh width of weakly crosslinked and fully swollen resins range between 10 to 100 angstrom units in size, (4,5).

The severely distorted geometry existing across chlor-alkali cell membranes operating under dynamic load can be treated using percolation theory. Percolation theory treats the degree of interconnectedness present in condensed matter in terms of a percolation transition that develops from increasing connectedness, identity and occupation within the network. The model accounts for the presence of a sharp phase transition at which long-range connectivity suddenly appears as a function of density, occupation and concentration and produces a second-order phase transition in the amorphous solid phase. Depicted in "Figure 5" is the concept of a critical volume fraction for percolation in the context of a two-dimensional honeycomb lattice. At the percolation transition point in the network, the underlying structure becomes topologically distorted and leads to anomalous selectivity in the thin permeable films.

During electrolysis ions within a membrane which is permeable to them do not remain permanently hydrated as the electric current drives them through the thin, ionically crosslinked structure of the microfine resin matrix. Movement of the hydrated cations toward the cathode occurs simultaneous with water stripping on the anodic side of the separator. The activity coefficient of water depends on molar concentrations of solutions on both sides of the separator. The transition from hydrated cations to cations with negative hydration displaces the activity versus concentration curve to higher molar values, as shown in "Figure 6". Data in this figure shows a typical functional plot of sodium transport number versus caustic and brine concentrations at $2KA/M^2$ and 90^oC with a perfluorosulfonic acid-carboxylic acid membrane system. The sulfonic acid portion is very much thicker than the carboxyl layer on the cathode side, (6). The existence of the maximum in the curve at the proper combination of all variables is thought to be due to a rapid viscosity increase at the critical molar concentration thereby generating a percolation transition zone at the cathode surface of the separator film. At this point the hydroxyl mobility decreases on the microscale within the micron thick cathodic membrane phase boundary corresponding to structural phase transformations within the constricted critical pore volume as pictured in "Figure 5".

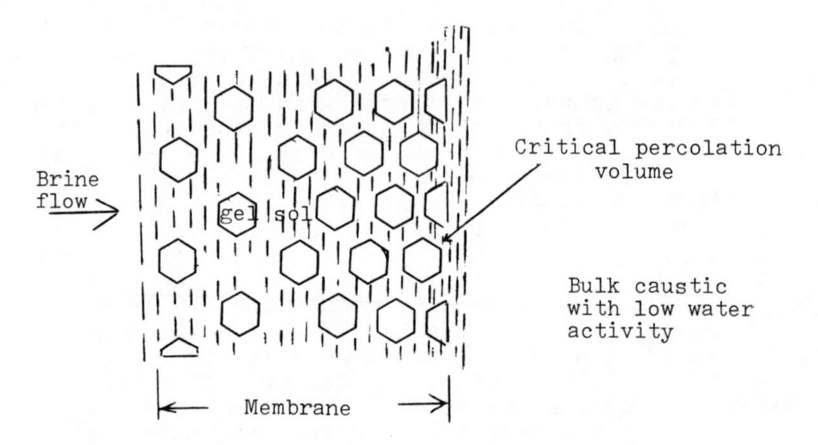

Figure 5. Critical volume flow diagram for the site-bond percolation about a gradient of gel-sol sites.

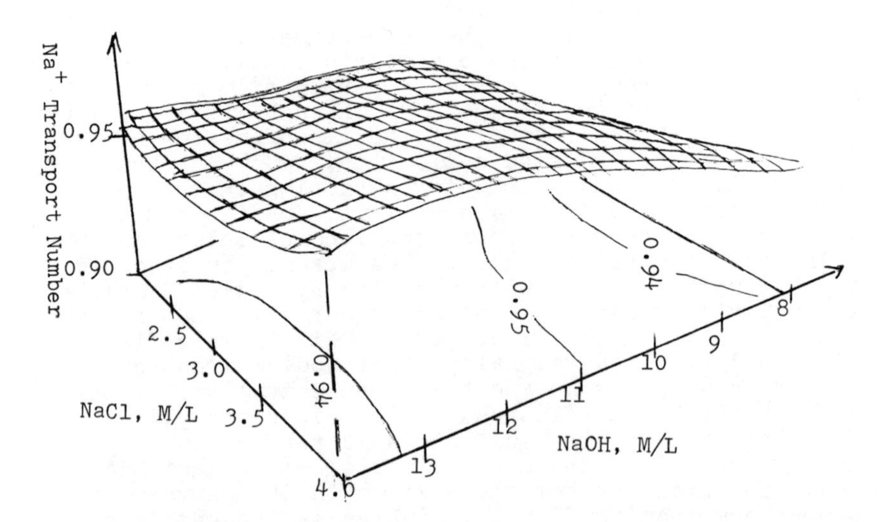

Figure 6. Three dimensional plot of sodium transport number versus anolyte-catholyte concentrations.

Only after viewing the membrane as a thin film semiconductive phase can one begin to seriously evaluate its potentialities. It is a multidimensional problem, and in the chlor-alkali cells the water transport is controlled by brine concentration while caustic strength controls the cathode efficiency. The membrane provides a low energy pathway for the phase change and separation process.

The flux of water, salt and ions through the membrane can be defined in a net flow equation. The net equation, neglecting convection and surface coverage, is:

$$dN/dt = (dN/dt)_D + (dN/dt)_O + (dN/dt)_{EM} + (dN/dt)_{ET} \qquad (2)$$

where:

$(dN/dt)_D$ = salt flux, and ion flux under diffusional control, $k_S(C_{S1}-C_{S2})$, mol/time-area.

$(dN/dt)_O$ = water flux under pressure control, osmotic flux, $k_w(\Delta P-\Delta\pi)$.

$(dN/dt)_{EM}$ = electromigration, $k_E(dE/dx)$.

$(dN/dt)_{ET}$ = charge transfer at electrodes, $k_{ET}(\exp[\alpha zF\eta/RT])$

The increased density of the thin cathodic barrier in a chlor-alkali membrane provides the second phase of matter required as the separating agent for alkali and chloride. This density change in the cathodic film selects a different, more constrictive innerchannel geometry at the critical percolation volume point, where selectivity is maximized as manifested by a maximum in the plot of sodium transport versus concentration, current, temperature and pressure. This density changes exponentially between points x_1 and x_2 in "Figure 7", along the thickness axis. This barrier film is less than 10 microns in depth but strongly modulates the intrinsic conduction as it increases voltage and the selectivity at the maximum in current efficiency.

Equation (2) is the base relationship electrochemical engineers utilize to describe and optimize the cells and make compromises among the competing factors such as: space-time yield, energy consumption, product quality and materials of construction.

Summary. Membrane cell processes have become important to modern technology to a great extent because of the development and utilization of perfluorinated membranes. The combination of metal anodes and the perfluorinated membranes has provided a modern revolution in the area of chlor-alkali production.

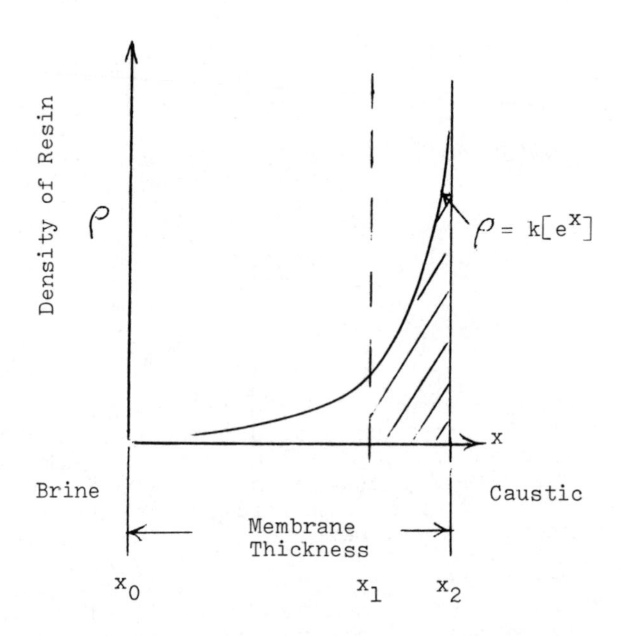

Figure 7. Density gradient across the polymer film with brine and caustic across the barrier.

These membranes are characterized by their structure and function with resin formulation based on the poly-tetrafluoroethylene backbone with short polyether side-chains terminating in sulfonic acid or carboxylic acid groups. These functional groups can join through ionic cross-links forming ion-clusters.

The strong interactions set up by densifying the thin cathodic barrier film on the membrane improve the degree of separation; thus the selectivity of these films to the extent that high efficiencies can be achieved at proper conditions of operation. One effective method of increasing this selectivity and throughput capacity of these films is to incorporate a substance that reacts chemically with one component transmitted, or that can interact strongly with it. The basic idea is to improve the sodium to caustic flux ratio by modification of the cathodic films, or adjust the caustic strength to operate at the minimum solubility point for caustic or potassium hydroxide, $NaOH3\frac{1}{2}H_2O$, or $KOH4H_2O$.

The topological distortion across these polymer films can be described by modern percolation theory, and used to define the maximum in the sodium transport versus caustic concentration curve as an optimum restricted pore volume.

Basic mathematical relations can be developed to treat the overall process in chlor-alkali production, as well as many of the other membrane tasks.

The process of "unmixing", or transforming a mixture of substances into two or more products that differ from each other in composition is served well by the membrane processes. This separation process is in stark contrast to natural forces, as Clausius so aptly put it: "Die entropie der welt strebt einem maximum zu".

Literature Cited

1. Dotson, R. L.; O'Leary, K. J., U.S. Patent 4,025,405.
2. Yeager, H. L.; Kipling, B.; Dotson, R. L., J. El. Chem. Soc., 127, No. 2, 1980.
3. Dotson, R. L.; Woodard, K. E., ACS Symposium Series, Washington, D.C., 1982.
4. Eisenberg, A.; King, M., "Ion-Containing Polymers, (Physical Properties and Structure)", Vol. 2, Academic: New York, 1977; Flory, P. J., "Principles of Polymer Chemistry", University Press: Ithaca, N.Y., 1953.
5. Helfferich, F., "Ion Exchange", McGraw-Hill, N.Y., 1962.
6. Kesting, R. E., "Synthetic Polymer Membranes", McGraw-Hill, N.Y., 1971.

RECEIVED June 10, 1985

11

The Chloralkali Electrolysis Process

Permselectivity and Conductance of Perfluorinated Ionomer Membranes

H. L. Yeager and J. D. Malinsky

Department of Chemistry, University of Calgary, Calgary, Alberta, Canada T2N 1N4

Perfluorinated ionomer membranes have been developed for use as separators in chlor-alkali electrolysis cells. Using an automated test apparatus, the current efficiency and voltage drop of such a high performance membrane were evaluated as a function of several cell parameters. Results are plotted as three dimensional surfaces, and are discussed in terms of current theories of membrane permselectivity.

The development of perfluorinated ionomer membranes for use in the production of chlorine and sodium hydroxide has proved to be a leading success in the field of membrane technology. In fact, the major industrial process of brine electrolysis, which has been employed for decades using both diaphragm and mercury cathode cells, has been revolutionized by these polymer membranes. The modern, high performance chlor-alkali membrane must possess the following capabilities: high physical strength and chemical stability, large ionic conductance, and low permeability to hydroxide ion - even when in contact with hot NaOH solutions of up to 15 M concentration. These performance goals have now largely been attained by continued improvements through several generations of materials. Currently, commercial perfluorinated ionomer materials for this application consist of membranes with carboxylate or mixed carboxylate-sulfonate functionality; the latter membranes often have layered structures with the carboxylate layer exposed to the caustic catholyte solution. Fabric reinforcement is used in some instances to improve strength.

While these membranes exhibit sodium ion transport numbers as high as 0.98 mol F^{-1} (i.e. only 2% of the electrolysis current is carried by hydroxide ion through the membrane) no comprehensive theoretical treatment of this unusually high permselectivity has yet emerged. The variation of permselectivity as a function of various cell parameters is also of interest, not only for practical reasons but also because of the insight that may be gained into the nature of hydroxide ion rejection. This research is directed at the latter problem, that is the characterization of membrane permselectivity

0097-6156/86/0302-0144$06.00/0

and resistance as a function of solution concentration, temperature, and current density.

Experimental

A laboratory membrane brine electrolysis cell, designed for automated operation, was constructed (1,2). This system enables the measurement of the sodium ion transport number of a membrane under specific sets of conditions using a radiotracer method. In such an experiment, the sodium chloride anolyte solution is doped with ^{22}Na radiotracer, a timed electrolysis is performed, and the fraction of current carried by sodium ion through the membrane is determined by the amount of radioactivity that has transferred to the sodium hydroxide catholyte solution. The voltage drop across the membrane during electrolysis is simultaneously measured, so that the overall performance of the material can be evaluated.

A block diagram of the apparatus is shown in Figure 1. The system is constructed to use three sodium chloride anolyte and four sodium hydroxide catholyte concentrations. The starred anolyte compartments refer to separate solutions which have been doped with radiotracer. These solutions are used only for determinations of transport number; the nonradioactive brine solutions are used for system flushing and membrane equilibrations. Solutions are selected and pumped into the cell, under computer control, through an all-Teflon pump-valve system. The solutions are heated during these transfers to ensure rapid attainment of experimental temperature in the cell. The brine system is designed to enable the return of radiotracer solutions to their storage vessels after each use. This serves to reduce consumption of radioactive solutions.

In practice, solutions are first pumped into the test cell, and then circulated for a period of several hours to condition the membrane. Next, an electrolysis is performed to further condition the material. These solutions are then discarded, fresh catholyte and radioactive anolyte are added, and electrolysis is conducted at a given membrane current density. A sample of anolyte and the entire catholyte solution are then pumped to a sample collector for weighing and determination of radioactivity. Other experiments may then be repeated at other current densities, or the sequence repeated with new solution concentrations. Thus twelve different combinations of anolyte and catholyte concentrations are used.

In this investigation, a sample of Nafion NX-90209 chlor-alkali membrane was used (E.I. du Pont de Nemours and Co., Polymer Products Department, Wilmington, DE). This membrane has sulfonate and carboxylate polymer layers and is reinforced with an open weave fabric.

Results and Discussion

Measurements were performed using 2, 3, and 4 M NaCl anolyte and 8, 10, 12, and 14 M NaOH catholyte solutions. The cell temperature was varied between 80° and 90°C and membrane current density was varied between 3 and 8 kA m^{-2} to test the effect of these parameters on membrane performance. For a given temperature and current density, values of t_{Na^+} were used to create a performance surface, using tensioned cubic spline functions (3,4). Surfaces for three different combinations of temperature and current density are shown in Figures 2-4. As seen for this (and other) high performance chlor-alkali membranes, sodium ion transport number shows a maximum with increasing caustic concentration but is much less sensitive to brine anolyte concentration. The position of this maximum changes with temperature and current density.

The very high values of t_{Na^+} (approaching 0.98 under certain conditions) and its variation with changes in cell parameters are subjects of great theoretical and practical interest. It should be noted that the lack of hydroxide ion migration through the membrane is not due to a Donnan exclusion process; considerable sorption of NaOH is found in these perfluorinated ionomer membranes when they are exposed to caustic solutions (5).

Therefore the high permselectivity seen for these membranes in concentrated caustic media must be attributed to a kinetic source. Several treatments of the transport properties of perfluorinated ionomers have been presented, using various approaches (6-12). Most of these have included the unusual ion clustering (13) of the polymer as a central morphological factor. Gierke's treatment suggested that channels which connect ionic clusters may serve as electrostatic barriers to anion transport. Using the Poisson-Boltzman equation to calculate the magnitude of this barrier, trends in observed current efficiency with polymer equivalent weight were obtained.

Reiss and Bassignana (8) note that such electrostatic barriers to anion transport would also serve as wells for cation transport, and that the transport of both cations and anions must be considered to explain what the authors term "superselectivity" of these polymers. Clearly though, the spatial variation in the concentration of fixed ion exchange sites within these ionomers is seen as the underlying cause of high permselectivity (8,12,14). Inherent in this view is the requirement that the symmetry of the variation is of a form that will affect cation and anion transport to differing degrees.

Datye and coworkers (12) treat an ionic cluster as a sphere with an electrical dipole layer at the surface. Ions of opposite charge would experience a different change in potential energy when traversing this dipole layer, giving rise to inherent permselectivity. This general approach looks very promising in terms of being able to generate a model which can eventually predict ionic transport properties from molecular and structural characteristics of these ionomers.

The variation of current efficiency with solution concentration in the chlor-alkali environment is an added complicating feature of these membranes' behavior. Kruissink (9) has performed elaborate calculations to yield the effect of electro-osmotic water transport on permselectivity, in classical terms. Results suggest that the minimum seen in t_{Na^+} (at lower NaOH concentrations than used here) may be due to the effects of electro-osmosis. Since hydroxide ion is the transported anion in chlor-alkali membrane cells, the possibility exists that non-classical transport involving proton tunneling events is also a contributing factor. Mauritz and Gray (15) have investigated the extent of proton tunneling events in a perfluorosulfonate film in contact with caustic solutions. They show

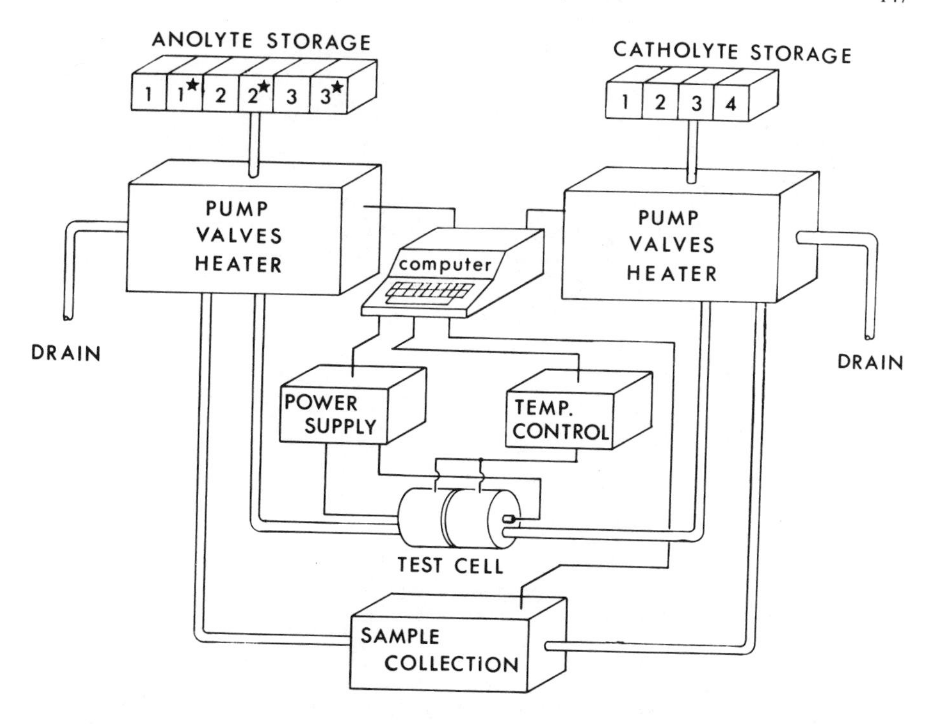

Figure 1. Block diagram of automated membrane test cell apparatus. Reproduced with permission from Ref. 1. Copyright 1982, the Electrochemical Society, Inc.

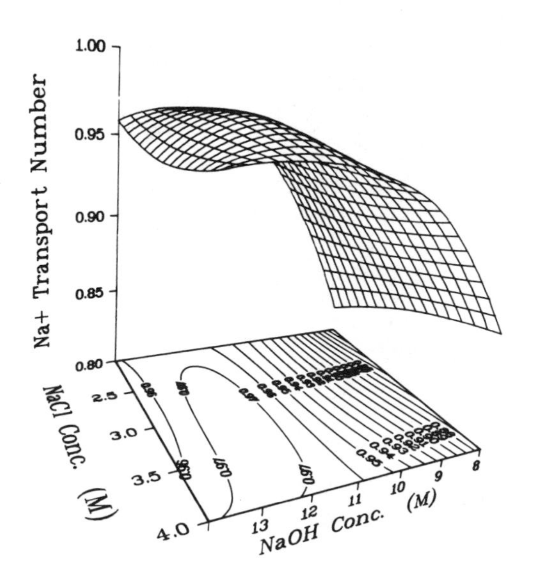

Figure 2. t_{Na+} for Nafion NX-90209, 90°C, 3 kA m^{-2}.

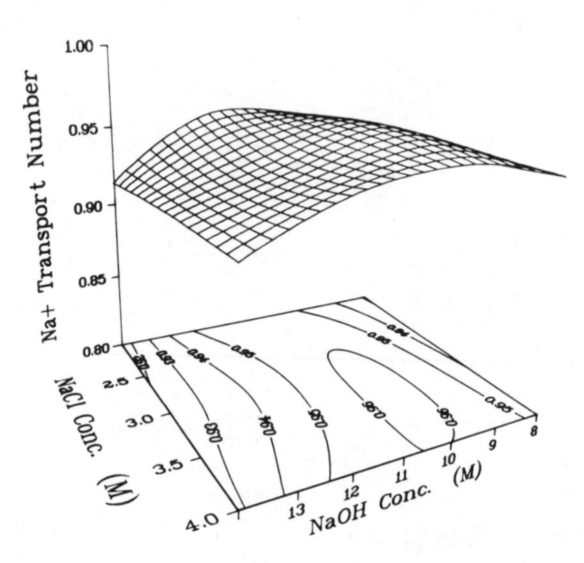

Figure 3. t_{Na^+} for Nafion NX-90209, 80°C, 3 kA m^{-2}.

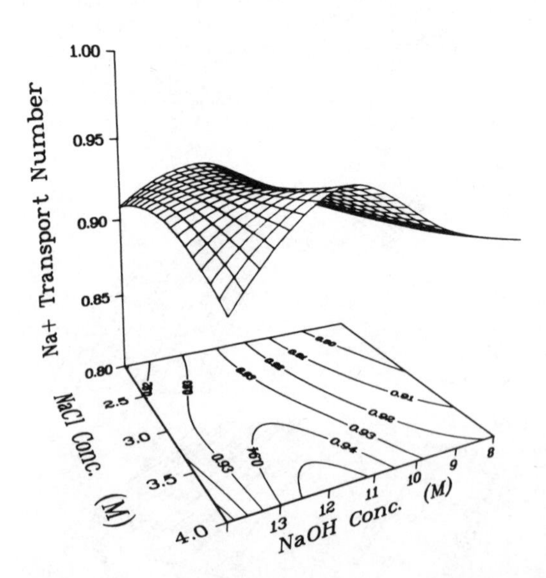

Figure 4. t_{Na^+} for Nafion NX-90209, 90°C, 8 kA m^{-2}.

remarkable parallels between an infrared continuous absorption of this polymer (reflecting proton tunneling) and its permselectivity in an operating chlor-alkali cell as a function of caustic catholyte concentration. The effect of increasing membrane dehydration with solution concentration on proton tunneling events (and hydroxide mobility) is seen as the likely cause of this relationship.

Thus, an overall theory of membrane permselectivity in terms of polymer properties may have to take into account a variety of factors to be a successful predictive tool in membrane design.

In practical terms, the variation in permselectivity as a function of various cell parameters means that cell performance must be optimized to minimize energy consumption. Not only membrane permselectivity but membrane voltage as well as other components of cell voltage must be considered in order to optimize cell power consumption per unit of product. Membrane voltage drop for Nafion NX-90209 is shown at 90°C and 3 kA m^{-2} current density in Figure 5, as a typical case. The membrane resistance increases monotonously with increasing caustic strength, probably due to increasing dehydration; in addition, the rise in voltage with increasing current is virtually ohmic.

The power consumption of a membrane chlor-alkali cell which uses this material can be estimated from these voltage and transport number results. Cell voltage is the sum of membrane voltage drop, galvanic voltage, electrode overpotentials, and solution and structural IR drops. When these are summed, cell power consumption is calculated by:

$$\text{Power Consumption (kWh tonne}^{-1}) = 670.1\ E_{cell}/t_{Na+}$$

Experimentally, we find that the galvanic voltage of the cell varies between 2.2 and 2.4 volts under these ranges of conditions. The other voltage components of the cell, for a finite gap configuration with low overvoltage anode and cathode, were estimated by the term

$$0.12 + 0.04\ (I + \ln I)\ \text{volts}$$

where I is current density in kA m^{-2}.

Power consumption trends, estimated in this manner, are shown in Figures 6-8 for three sets of operating conditions. At 90°C and 3 kA m^{-2} current density, a broad minimum is seen in power consumption with increasing caustic concentration. At 80°C, power consumption is lowered for more dilute caustic concentrations due to the shift of the maximum in t_{Na+}. At 90°C and 8 kA m^{-2}, power consumption rises about 20-30% compared to results for 3 kA m^{-2} at the same temperature. Values are greatest at higher caustic strengths, a result of both large membrane voltages and reduced membrane current efficiencies. Overall, cell power consumption changes slowly and monotonically with these cell parameters for this high performance membrane.

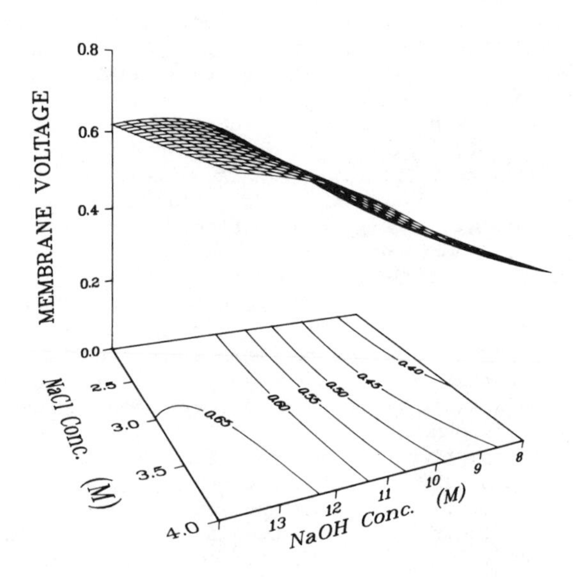

Figure 5. Membrane voltage drop, 90°C, 3 kA m^{-2}.

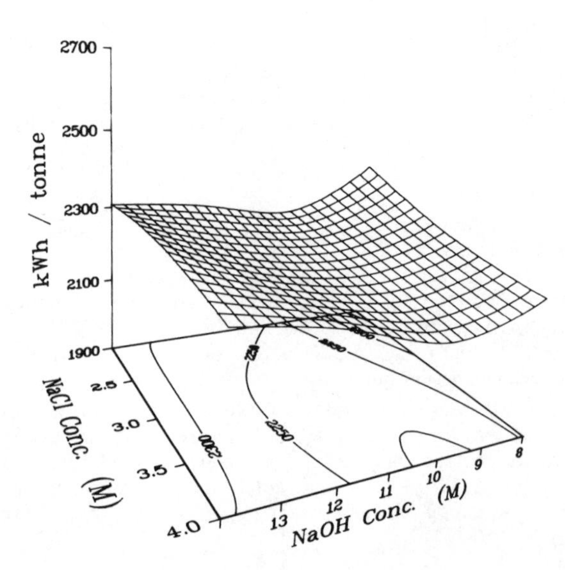

Figure 6. Estimated cell power consumption, 90°C, 3 kA m^{-2}.

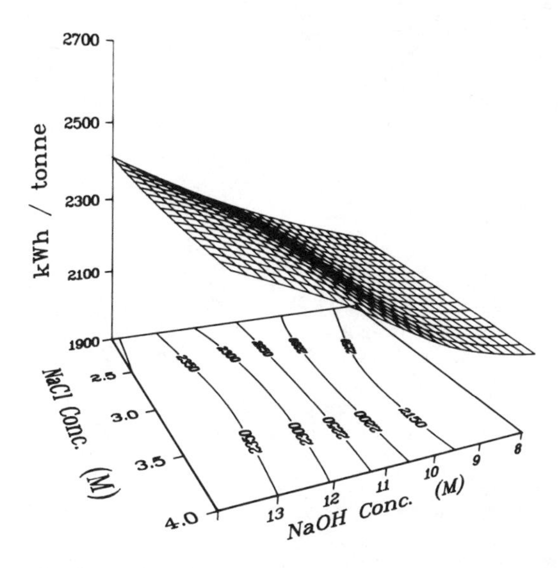

Figure 7. Estimated cell power consumption, 80°C, 3 kA m^{-2}.

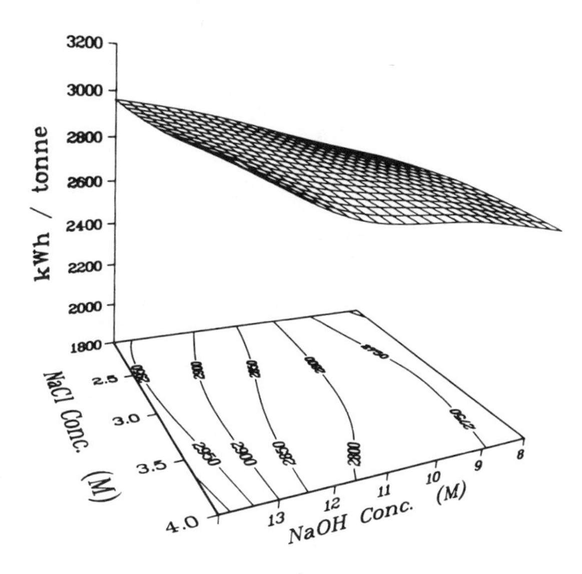

Figure 8. Estimated cell power consumption, 90°C, 8 kA m^{-2}.

Literature Cited

1. H.L. Yeager, J.D. Malinsky, and R.L. Dotson, Proc. Sym. Transport Processes Electrochem. Sys. The Electrochemical Society, Inc., Pennington, N.J., 1982, 215.
2. H.L. Yeager, in "Proc. Sym. Mem. and Ionic and Elec. Cond. Poly." The Electrochemical Society, Inc., Pennington, N.J., 1982, 134.
3. A.K. Cline, Comm. of Assoc. Computer Machinery 1974, 17, 218.
4. A.K. Cline, Comm. of Assoc. Computer Machinery 1974, 17, 220.
5. Z. Twardowski, H.L. Yeager, and B. O'Dell, J. Electrochem. Soc. 1982, 129, 328.
6. T.D. Gierke, Paper 438 presented at the Electrochemical Society Meeting, Atlanta, GA., Oct. 9-14, 1977.
7. W.S. Hsu, J.R. Barkley, and P. Meakin, Macromolecules 1980, 13, 198.
8. H. Reiss and I.C. Bassignana, J. Mem. Sci. 1982, 11, 219.
9. Ch. A. Kruissink, ibid. 1983, 14, 331.
10. W.H. Koh and H.P. Silverman, ibid. 1983, 13, 279.
11. W.S. Hsu and T.D. Gierke, ibid. 1983, 13, 307.
12. V.K. Datye, P.C. Taylor, and A.J. Hopfinger, Macromolecules 1984, 17, 1704.
13. "Perfluorinated Ionomer Membranes", A. Eisenberg and H.L. Yeager, Eds., American Chemical Society, Washington, D.C., 1982.
14. C. Selvey and H. Reiss, in press.
15. K.A. Mauritz and C.L. Gray, Macromolecules 1983, 16, 1279.

RECEIVED June 10, 1985

12

Solution Processing of Perfluorinated Ionomers

Recent Developments

Michael J. Covitch[1]

Diamond Shamrock Corporation, Painesville, OH 44077

Several new solvents for sulfonate and carboxylate perfluorinated ionomers such as Nafion (E. I. duPont de Nemours and Company) give rise to a number of interesting and technologically significant applications of these materials. The ability of a particular solvent to dissolve 1100-1200 equivalent weight ionomers is influenced by the type of bound anion as well as the cation chemistry. For example, N-butylacetamide dissolves both sulfonate and carboxylate ionomers; and it can be used to prepare ionomer alloys having mixed functionality. Porous structures can be prepared from a solution of the ionomer precursor ($-SO_2F$ or $-COOCH_3$) in a perfluoroalkylacid solvent by pseudo eutectic solidifcation techniques. Applications of these dissolution processes for the preparation of membranes, solid polymer electrolytes, and coatings are briefly discussed.

Perfluorinated ionomers such as Nafion are of significant commercial importance as cation exchange membranes in brine electrolysis cells (1). Outstanding chemical and thermal stability make this class of polymers uniquely suited for use in such harsh oxidizing environments. The Nafion polymer consists of a perfluorinated backbone and perfluoroalkylether sidechains which are terminated with sulfonic acid and/or carboxylic acid functionality.

$$\left(CF_2CF_2\right)_x\left(CF_2\underset{\underset{\underset{\displaystyle (CF_2\underset{\displaystyle CF_3}{CF}\text{-}O)_z(CF_2)_n X}{|}}{O}}{|}}{CF}\right)_y$$

where X = SO_3M, CO_2M; M = H, metal cation; $Z \geq 1$; n = 1,2.

[1]Current address: The Lubrizol Corporation, Wickliffe, OH 44092

0097–6156/86/0302–0153$06.00/0

In a previous paper (2), the author described a method to dissolve the sulfonyl fluoride precursor form of a perfluorinated sulfonate ionomer. Commercially available forms of Nafion are supplied as activated membranes (i.e., saponified from the precursor to the ionic form), and near-quantitative reconstitution of the precursor functionality (such as RSO_2F) must first be performed using a chemical reagent such as SF_4 (4) before dissolution in perhalogenated solvents is possible. Besides adding to the cost of membrane manufacture, SF_4 is extremely toxic and corrosive and must be handled in nickel alloy pressure equipment. Therefore, a method for dissolving perfluorinated ionomers directly would be more desirable.

Martin and co-workers (5) published a procedure for solubilization of perfluorinated ionomers in alcohol/water solutions using a combination of heat and pressure. Although Martin (18) has claimed that only low concentration solutions (<1% w/v) are stable at atmospheric pressure, W. G. Grot (19) has obtained stable, one-phase solutions at concentrations greater than 5 weight %. Oyama, et al (20) reported that a dilute solution of Nafion 125 (1200 equivalent weight, sodium salt) can be obtained by boiling the polymer in dimethylsulfoxide (boiling point = 189°C) followed by filtration. A 5.6 mg/ml (0.56% w/v) stock solution was prepared by 1:1 (v/v) dilution of the filtrate with ethanol and used for coating graphite electrodes. The present paper describes a number of other solvents which are capable of dissolving perfluorinated ionomers (up to ca. 10 wt. % solids) at atmospheric pressure and elevated temperature.

Experimental

The 1100 equivalent weight sulfonate ionomer was prepared from Nafion 117 by ion exchange of the sulfonic acid form (prepared according to reference 2) with the appropriate metal hydroxide. Carboxylate ionomer was available as the catholyte facing layer of Nafion 901. This material was converted to the acid form by three successive four-hour treatments in pH > 3 aqueous HCl, maintaining the pH as close to three as possible by dropwise addition of 1 wt. % HCl. This procedure is necessary to maintain ion pair dissociation of the carboxylic acid group. Ion pair association begins to occur at external solution pH < 3 which causes the polymer to deswell and become milky white due to the inclusion of entrapped acid. The metal carboxylate was then prepared by the same method described above for the sulfonate ionomers. The procedure for preparation of the sulfonyl fluoride form of Nafion 117 has been outlined elsewhere (4). The carboxylic acid form of Nafion 901 was esterified in methanol by bubbling dry HCl gas through the solution at 20-30°C for four hours. All samples were washed with distilled water and dried under vacuum at 80°C for 24 hours.

All solvents were purchased commercially and used without further purification. Ionomer dissolution was accomplished with mechanical agitation under a blanket of dry nitrogen at approximately 10°C below the boiling point (at atmospheric pressure) of the solvent but no higher than 230°C.

Results and Discussion

Besides the perhalogenated compounds which successfully dissolve the sulfonyl fluoride and methyl carboxylate ionomer precursors (2), a number of polar non-halogenated organics (see Table I) have been identified which dissolve either the sulfonate or the carboxylate ionomer (6). In general, the smaller alkali metal (such as Li^+, Na^+) forms are more readily dissolved than those containing larger cations. Solutions containing 8-10% by weight of ionomer can be readily prepared from most of the solvents listed in Table I. In contrast to the precursor solutions in halogenated solvents, the ionomer solutions are far less viscous and do not become gelatinous when cooled to 30°C. In the 20-30°C temperature regime, certain solutions such as those in Sulfolane (Phillips Petroleum) begin to thicken. Therefore, these ionomer solutions are easier to prepare and more convenient to handle than the previously described precursor solutions.

One of the solvents on Table I--N-butylacetamide--dissolves both the sulfonate and carboxylate layers of Nafion 901; sulfolane cleanly dissolves away only the sulfonate layer, leaving an intact carboxylate film. One of the most intriguing applications of these dissolution methods is in the preparation of novel sulfonate/carboxylate ionomer alloys. The patent literature has described the usefulness of perfluorinated ionomer alloy membranes for brine electrolysis (7), and a recent paper by Hsu and co-workers (8) describes the effects of alloy morphology on membrane selectivity. These alloys were produced by extrusion, adjusting the melt temperature and presumably the film take-up speed to control morphology. Utilizing N-butylacetamide as a cosolvent, ionomer alloy membranes may also be prepared by standard film coating techniques. Here, control of alloy morphology may be exercised by adjusting the blend composition, casting temperature, and drying rate. In addition, mixed solvents may be used to preferentially extend one domain at the expense of the other, thereby adding another dimension to domain morphology control.

If the precursor forms of the sulfonate and carboxylate ionomers are available, alloys may also be prepared from solutions in Halocarbon Oil (Halocarbon Products Corporation).

In a previous publication (2) it was shown that solutions of the sulfonyl fluoride precursor in certain perfluorinated solvents such as perfluorooctanoic acid "form solids at room temperature . . . by virtue of the melting points of their respective solvents." Removal of the solvent by dissolution or vacuum sublimation leaves a porous structure (6), the pores having been formed by crystallization of solvent which separated by syneresis from the gel during cooling. The porous product is best saponified in aqueous metal hydroxide solution prior to solvent removal to stabilize the pore structure. Similar results have been obtained by others (9-13) for polyolefins dissolved in high melting diluents. In these references it is noted that the two-phase morphology of a quenched polymer solution is related to the overall composition, cooling rate, and quenching temperature;

Table I. Solvents capable of dissolving perfluorinated ionomers and ionomer precursors at atmospheric pressure and elevated temperature.

Solvent	Functional Group				Boiling Point °C
	SO_2F	$CO_2^-Z^+$	CO_2CH_3	$SO_3^-Z^+$	
Halocarbon Oil	X		X		225-260
perfluorooctanoic acid	X		X		189
perfluorodecanoic acid	X		X		218
perfluorotributylamine	X				175
FC-70 available from 3M (perfluorotrialkylamine)	X				215
perfluoro-1-methyldecalin	X				159
decafluorobiphenyl	X				206
pentafluorophenol	X				143
pentafluorobenzoic acid	X				220
N-butylacetamide		X		X	229
sulfolane (tetramethylene sulfone)				X	285
N,N-dimethylacetamide				X	165
N,N-diethylacetamide				X	185
N,N-dimethylpropionamide				X	174
N,N-dibutylformamide				X	242
N,N-dipropylacetamide				X	209
N,N-dimethylformamide				X	153
1-methyl-2-pyrrolidinone				X	202
diethylene glycol				X	245
ethylacetamidoacetate				X	265

Z is an alkali or alkaline earth metal or a quaternary ammonium ion having attached hydrogen, alkyl, substituted alkyl, aromatic, or cyclic hydrocarbon.

the smallest pores are generally formed by solidification of the eutectic composition. Photo-micrographs of quenched perfluorosulfonyl fluoride/perfluorooctanoic acid solutions and those found in reference 9 are quite similar. Electron micrographs and simple permeation experiments confirm the porous nature of the solvent extracted perfluorinated ionomer. Porous membranes of this type may find application in chlor-alkali diaphram cells and solid polymer electrolyte electrodes (6).

Perfluorinated ionomer solutions may be applied to a variety of substrates to form fabric reinforced membranes, solid polymer electrolytes, and carboxylate-coated sulfonate membranes (6,14). Solvent cast films are as durable as extruded membranes of the same thickness and display identical ion selectivity characteristics. Current membranes are often reinforced with a fabric to reduce the possibility of tearing during handling. The fabric is sandwiched between the membrane (in the precursor thermoplastic form) and a release paper and impregnated into the membrane on a heated vacuum roll fitted with a concentric horseshoe heater (16). This process does not completely encapsulate the fabric, resulting in exposed fabric "knuckles" on the release paper side of the membrane. In addition, the fiber-like topography of the release paper is transferred to one side of the membrane. This rough membrane surface may affect electrolyte flow patterns past the membrane, resulting in stagnant pockets in which severe electrolyte depletion or concentration may occur. Concentration polarization effects increase leading to an increase in cell voltage. By comparison, solution coated fabric is totally encapsulated by the membrane, and both membrane surfaces are smooth. Thus, electrolyte flow patterns past both surfaces are similar, avoiding unusual concentration polarization affects at one surface.

A solid polymer electrolyte (SPE) consists of a membrane which is in direct contact with both electrodes, thereby eliminating electrolyte gap resistance to reduce all voltage. The General Electric SPE (17) consists of two porous particulate electrodes which are bonded cohesively with polytetrafluoroethylene dispersion particles and connected electrically to the outside of the cell hardware by means of metallic current collectors which are pressed against the SPE by mechanical methods. Such an SPE can be prepared via perfluoroionomer solution techniques. One method is to apply a paste consisting of the electrolyte powder and the perfluoroionomer solution to the membrane and evaporate the solvent. Alternately, the paste can be applied to a sacrificial substrate such as aluminum foil, dried, and subsequently pressed into the membrane as a decal. The use of a perfluorinated ionomer as the SPE electrode binder results in better adhesive bonding with the membrane by virtue of the fact that the binder and the membrane are of identical composition. In addition, the solvent in the electrode paste promotes "solvent welding" by softening the membrane during the application process.

Perfluoroionomers may also be applied via the solution process as a protective coating to reaction vessels or other metallic equipment (15) to prevent corrosion or product build-up.

Corrosion resistant fluoropolymer coatings are currently marketed by W. L. Gore & Associates, Inc. (Fluoroshield) and Pfaudler.

Acknowledgments

The author wishes to thank Drs. G. H. McCain and L. L. Benezra for their valuable suggestions and criticism; Mr. G. G. Sweetapple for his technical assistance; and to The Lubrizol Corporation for its assistance in the preparation and presentation of the manuscript.

Literature Cited

(1) W. G. Grot, G. E. Munn, and P. N. Walmsley, Paper 154 presented at the Electrochemical Society Meeting, Houston, TX, May 7-11, 1972.
(2) G. H. McCain and M. J. Covitch, J. Electrochem. Soc., 1984, 131(6), 1350.
(3) E. I. duPont de Nemours & Company, "Nafion Perfluorinated Membranes" product literature, February 1, 1984.
(4) M. J. Covitch, U. S. Patent 4,366,262 (1982).
(5) C. R. Martin, T. A. Rhoades, and J. A. Ferguson, Anal. Chem., 1982, 54, 1641.
(6) M. J. Covitch, D. L. DeRespiris, L. L. Benezra, and E. M. Vauss, U.S. Patent 4,421,579 (1983).
(7) C. J. Molnar, E. H. Price, and P. R. Resnick, U. S. Patent 4,176,215 (1979).
(8) W. Y. Hsu, T. D. Gierke, and C. J. Molnar, Macromolecules, 1983, 16, 1947.
(9) P. Smith and A. J. Pennings, Polymer, 1974, 15, 413.
(10) E. Calahorra, G. M. Guzman, and F. Zamora, J. Polymer Sci.: Polymer Lett. Ed., 1982, 20, 181.
(11) M. J. Covitch, Eur. Pat. Appl. EP 69,516 (1983), see Chem. Abstr. 98, 127309m.
(12) P. Smith and A. J. Pennings, J. Materials Sci., 1976, 11, 1450.
(13) P. Smith and A. J. Pennings, J. Polym. Sci.: Polym. Phys. Ed., 1977, 15, 523.
(14) M. J. Covitch, M. F. Smith, and L. L. Benezra, U. S. Patent 4,386,987 (1983).
(15) S. K. Baczek, G. H. McCain, and M. J. Covitch, U.S. Patent 4, 391, 844 (1983).
(16) D. E. Maloney, U. S. Patent 4, 349, 422 (1982).
(17) T. G. Coker, R. M. Dempsey, adn A. B. La Conti, U.S. Patent 4, 210, 501 (1980) and U. S. Patent 4, 191, 618 (1980)
(18) C. R. Martin, private communication.
(19) W. G. Grot, private communcation. 5% solutions of 1000 or 1100 EW sulfonic acid resin in a mixture of lower aliphatic alcohols and water are commercially available from Solution Technology, Inc., Box 171, Menden Hall, PA 19357.
(20) N. Oyama, N. Oki, H. Ohno, Y. Ohnuki, H. Matsuda, and E. Tsuchida, J. Phys. Chem., 1983, 87, 3642.

RECEIVED June 10, 1985

13

Microstructure of Organic Ionic Membranes

M. Pinéri

Groupe de Physico-Chimie Moléculaire, Département de Recherche Fondamentale, Service de Physique, Centre d'Etudes Nucléaires de Grenoble, 85 X, 38041 Grenoble Cédex, France

Perm selectivity and ionic conductivity are the two important parameters, besides the mechanical and chemical stability, to be considered in applications for ion exchange membranes; no connection is usually done with the microstructure. The aim of this paper is to present results on the physical structure of perfluorinated membranes. Crystallinity and distribution of the ionic sites across the thickness will be first considered. Evidence of a separate ionic phase will then be given and the local ion concentration will be analyzed.

The physical structure of ion-containing polymers has been the subject of many investigations. Each of these polymers has a hydrogenated or fluorinated backbone with a statistical distribution along the chain of the ion exchange side groups. The percentage of monomers containing the ionic groups is usually lower than a. 10%; at larger concentrations the copolymers may become water soluble. Exchange of the carboxylic or sulfonic acid protons with other cations allows the use of a large range of spectroscopic techniques to better understand the microstructure of these materials. Condensation of charges has been theoretically proposed and also experimentally observed. The basic ion pairs $A^{-}X^{+}$ are first associated to form the so-called multiplets, which contain a small number of these ion pairs. Electric interactions are responsible for these aggregations and steric hindrance limits the extent of association. Clustering of multiplets may occur because of residual electric interactions between multiplets. The physical crosslinking gives some interesting properties to these materials.

Polymeric ion exchange membranes are a particular class of these ion-containing polymers. We have been interested in understanding their microstructure in order to explain membrane characteristics such as permselectivity, ionic conductivity, and water diffusion. Most of the results obtained on perfluorinated ionomer membranes are summarized in a book (1).

In this paper, we will report results mainly obtained on perfluorinated sulfonated membranes. First, we will show that the distribution of the ion exchange groups may be quite nonuniform on

0097-6156/86/0302-0159$06.00/0

both a macroscopic and a microscopic level. We will then concentrate on the so-called ionic regions for which we will define the chemical composition. The geometry of these domains will then be discussed.

Crystallinity

In terms of structure of the membranes the first point to be discussed concerns the possibility of having crystalline domains and the role of these domains. X ray and differential scanning calorimetry measurements are the best ways to probe the existence and the structure of crystalline domains. Figure 1 represents different X ray scattering curves obtained for:

- a low density polyethylene after irradiation grafting with styrene followed by sulfonation (1.46 meq/g exchange capacity) (figure 1a);
- a fluorinated polyvinylidene on which have been irradiation grafted monomers of dimethyl amino ethyl methacrylate followed by quaternization (figure 1b). This membrane is anionic, the formula is:

$$-(CH_2-CF_2)_n-CH_2-\underset{\underset{\underset{O}{\|}}{C-O-(CH_2)_2-\overset{+}{N}H(CH_3)_2\ OH^-}}{|}}{\overset{\overset{CH_3}{|}}{C}}$$

Indexes of crystallinity can be obtained from these different experiments. It has to be noted that the crystalline structure generally corresponds to the crystalline structure of the starting homopolymers. The crystalline domains play a role as physical crosslinks.

A similar behavior has also been observed in membranes obtained by copolymerization. Many recent perfluorinated sulfonated membranes have been obtained by copolymerization of tetrafluoroethylene with a sulfonyl fluoride vinyl ether. This high molecular weight polymer has the following formula:

$$-(CF_2-CF_2)_n-CF_2-\underset{|}{CF}-O-CF_2-\underset{\underset{CF_3}{|}}{CF}O-(CF_2)_2-SO_2F$$

In such a form this material is melt fabricable and after hydrolysis is converted to a ion exchange membrane with a perfluorosulfonate group, $-SO_3Na$. The sodium counter ion can be exchanged by other metal ion or hydrogen ion.

Figure 2 represents the neutron diffraction spectrum obtained with a copolymer corresponding to $n \sim 7$. The existence of crystalline domains with the same structure as for PTFE implies the absence of ion exchange groups in these domains. Such a result involves the presence along the backbone of many long segments containing only CF_2-CF_2. After swelling the broad peak is significantly broader for the wet sample than for the dry sample; this observation means that the wet material is more disordered and that the separation between amorphous and crystalline regions is less clear cut. This result implies that the crystallites which act as physical crosslinks are

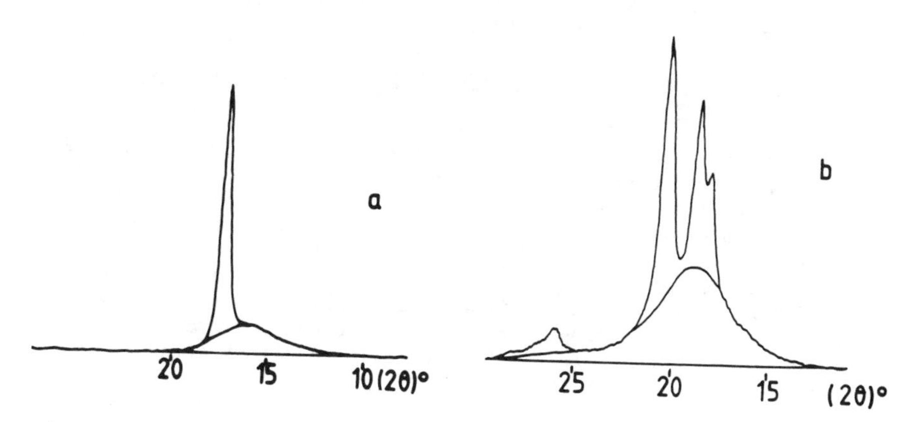

Figure 1. X-ray scattering spectra of membranes obtained from polyethylene (a) and polyvinylidene fluoride (b).

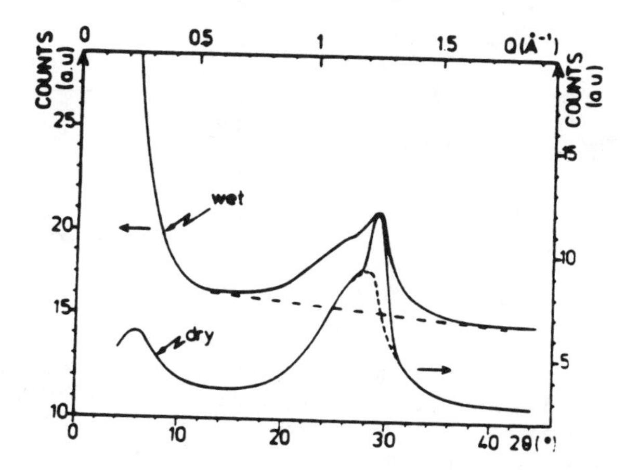

Figure 2. Neutron diffraction spectra of acid Nafion membrane for wet and dry samples. Instrument is D 1B. Incident wave-length is λ_0 = 2.56 Å. The horizontal axis is expressed either in scattering angles (2θ) units, or in neutron momentum transfer ($Q = 4\pi/\lambda_0 \sin\theta$) units. Note the different vertical scales for the two samples. The separation between the amorphous and crystalline contributions to the large angle peak is shown for the dry sample. Temperature ∿ 25°C. Reproduced with permission from Ref. 3. Copyright 1982 J. Polym. Sci., Polym. Phys.

partially destroyed by the chain tensions due to swelling of the ionic hydrophilic phase.

These crystallites therefore form a separate phase and do not participate in the exchange process.

Macroscopic Distribution of the Ion Exchange Sites across the Membrane Thickness

An electron microprobe is used to define the extent of the homogeneity for the distribution of the exchange sites across the thickness of the membrane. Cationic membranes are exchanged with Cu^{++}. Chlorhydration of quaternized N of anionic membranes results from the presence of a Cl ion per ionic site. The sample is bombarded with a beam of electrons and the number of emitted X rays in a certain energy range (corresponding to the Cu and Cl transitions) is counted. The number observed is proportional to the relevant elemental concentration. About 1 μm^3 of sample is probed. The electron gun and X ray detector are simultaneously swept at constant speed across the edge (∿ 250 μm thick) of a cut membrane giving a profile of elemental concentration versus distance into the membrane.

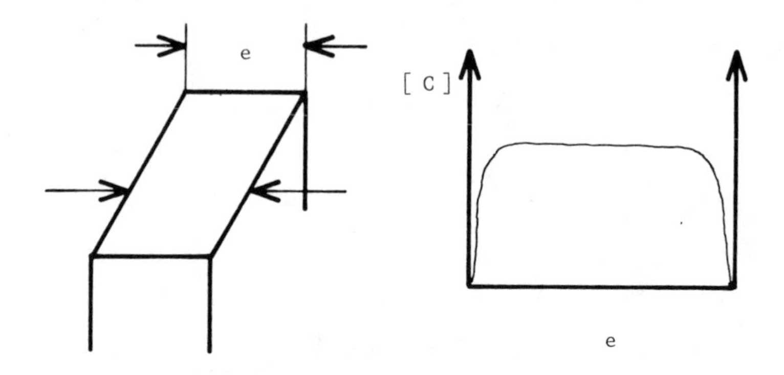

Concentration profiles obtained with different membranes are shown in Figure 3. Most of these membranes obtained by irradiation grafting have a nonhomogeneous distribution of ionic sites across the thickness. The exchange properties and the ionic conductivity must strongly depend on this asymmetry in the repartition. A new approach to asymmetric membranes may be envisaged from this irradiation grafting technique.

On the other hand, a homogeneous distribution is obtained for the Nafion perfluorinated membranes obtained by copolymerization and hydrolysis of the SO_2F side groups.

Let us now concentrate on these last membranes, which have a relatively homogeneous ionic site distribution at a μm scale. Nafion 1200 equivalent weights have been studied here. We already know that there are small microcrystallites with a PTFE structure. The question now concerns the distribution of SO_3H outside of these crystals; do we have a statistical distribution of them or do we have some ionic clusters with a larger ionic concentration? This question will be developed in the next chapter.

Microscopic Distribution of the Ionic Sites

Information concerning phase separation inside a material can be obtained from small angle X ray or neutron scattering. We have experimented with melt quenched perfluorinated membranes to get rid of the crystalline phase (2). Water absorption at room temperature is similar to that in the nonquenched material. Water has been shown to be absorbed specifically where the $SO_3^-H^+$ groups are in the acid form; the water tends to first surround the cation. Water protons constitute an excellent probe in neutron scattering because of the large differences in scattering length compared with C, F, S and O. With small angle neutron scattering (SANS), it is therefore possible to determine if there is a statistical distribution of the hydrated ionic groups or a clustering of them.

Figure 4 shows the SANS curve obtained for a Na^+ exchanged Nafion 1200 specimen first quenched from the melt and then boiled in water. The water content for this sample is around 20% by weight or 40% by volume. The existence of a peak at low Q ($Q = \frac{4\pi}{\lambda}\sin\theta$) and a zero order scattering increase in the first evidence of a nonuniform distribution of water inside the specimen. Similar curves have been obtained with samples containing less water (3). When the hydration level is decreased, we observed a decrease in the intensity of the peak and a shift at higher angles of this peak. The qualitative result obtained from these curves is the existence of ionic hydrophilic domains whose size and distance may depend on the total water content.

A more quantitative analysis of the extent of phase separation may be obtained by using the technique of isotopic replacement. Two samples are prepared having the same structure except that the atoms of an element found in one sample have been replaced by an isotope in the other. In this work, this is accomplished by hydrating samples with mixtures of H_2O and D_2O of various proportions. For samples that have two phase structures the ratio of SANS intensities is given by:

$$\frac{I'(Q)}{I(Q)} = \frac{\left[\bar{\beta}_1 - (\rho_2'/\rho_1'\ \beta_2')\right]^2}{\left[\beta_1 - (\rho_2/\rho_1)\beta_2\right]^2}$$

where primed quantities are those after isotopic replacement. The intensity ratio is independent of the scattering vector and the two scattering curves differ only by a constant multiple. For systems involving more than two phases this will not generally be the case.

Scattering curves have been obtained for samples containing different water amounts and a different H_2O/D_2O ratio. For example, for a specimen soaked in a mixture of 12.5% H_2O/87.5% D_2O (total water content is 16% by weight) essentially no scattering was observed above a constant background. This result in itself indicates that the scattering system contains essentially two phases or contrast regions. The matching concentration, known as the isopicnic point, corresponds to the concentration at which

$$\beta_1 = (\rho_2'/\rho_1')\beta_2$$

Since each value of β changes linearly with the ratio $[D_2O]/[H_2O]$ and the density values change little, $[I/I_{H_2O}]^{1/2}$ should be linear as

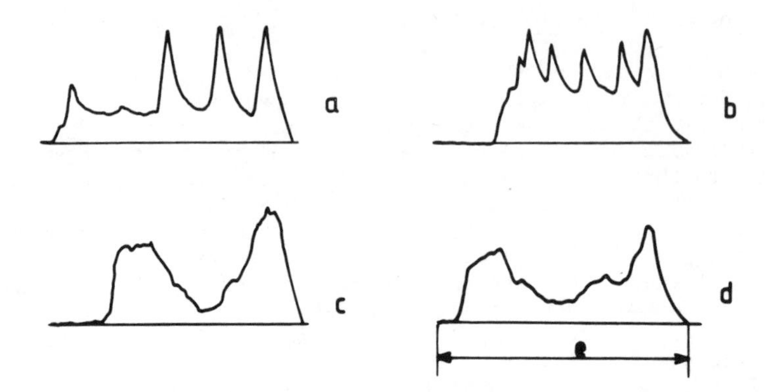

Figure 3. Concentration profiles for different membranes giving an element concentration (in arbitrary units) across the thickness a and b, Cu concentration after exchange. The membranes have been obtained from a low density polyethylene after irradiation grafting with styrene followed by sulfonation. c, Cu concentration after exchange. The membrane has been obtained from a copolymer of tetrafluororethylene and fluorinated propylene after irradiation grafting with styrene followed by sulfonation. d, Cl concentration after chlorhydratation of dimethyl annio ethyl methacrylate grafted on polyvinylidene fluoride.

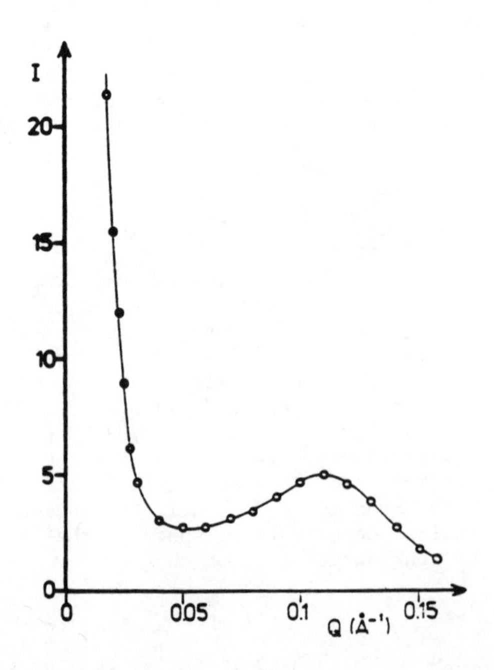

Figure 4. Characteristic curve obtained in a small angle scattering experiment showing both the peak and the zero order scattering increase.

a function of $[D_2O]/[H_2O]$. Such a dependence is shown in Figure 5 in which the constant of multiplication $[I/I_{H_2O}]^{1/2}$ experimentally obtained to give the best possible overall superposition is plotted versus $[D_2O]/[H_2O]$ for different water contents. A good linear dependence is observed for the high water content samples. For low water contents the two phase approximation is less valid. SAXS results confirm such a result with a positive deviation from Porod's law characteristic of density fluctuations within phases.

These SANS results therefore confirm the existence of a separated phase containing most of the ionic exchange sites and the hydration water.

Local concentration of cation and chemical composition of this ionic phase in different ion exchange ionomer membranes has been obtained from electron spin resonance (ESR) measurements of Cu^{++} exchanged cationic membranes (4). SANS experiments have shown that no drastic change occurs in the microstructure of these membranes upon neutralization. General information about this microstructure can therefore be obtained through use of cationic probes. A different situation exists in other ion containing polymers where a complete reorganization of the microstructure occurs upon neutralization of the acidic groups.

Local Cu^{++} concentrations in the membranes have been obtained by comparing the corresponding ESR spectra with the spectra obtained with reference Cu^{++} solutions. Frozen reference solutions (at liquid nitrogen temperature) permit us to get rid of the molecular motions. Prior to the experiment the samples have therefore been quenched from room temperature down to 77 K in the finger dewar of the ESR spectrometer. The reference solutions are aqueous solutions to which we have added a cryoprotector. It has been shown that such an addition prevents crystallization of water during cooling, thus preserving the total Cu^{++} concentrations. The Cu^{++} concentrations range of the reference solutions (0.02-0.2 g Cu^{++}/cm^3) has been chosen in order to match the possible concentrations found in the membranes. It turns out that in this concentration range there are dramatic changes in the shape of the ESR spectra, these changes are due to relatively important changes in the dipole-dipole and exchange interactions between the electronic spins. For low concentrations, the spectrum is very similar to that of isolated Cu^{++} ions where the four lines due to hyperfine interactions with the Cu^{++} nuclear spins are clearly seen. For higher concentrations only a broad line is seen. A continuous change in the spectra is observed when the concentration is varied (Figure 6). A few empirical parameters, such as the widths at half and quarter of maximum and relative heights of various components, have been used to characterize the spectra; the changes in these parameters vs. Cu^{++} concentrations have been plotted in Figure 7.

The various reference solutions obtained by changing both the nature of the salt (sulfate, nitrate, and bromure) and the cryoprotector (dimethylsulfoxide, glycerol, ethyleneglycol) have given essentially the same results for the same Cu^{++} concentrations. The changes observed from one spectrum to another are small compared to the changes observed with a small modification of the Cu^{++} concentration. We can therefore assume, in first approximation, that an ESR spectrum is characteristic of the Cu^{++} concentration. Because the

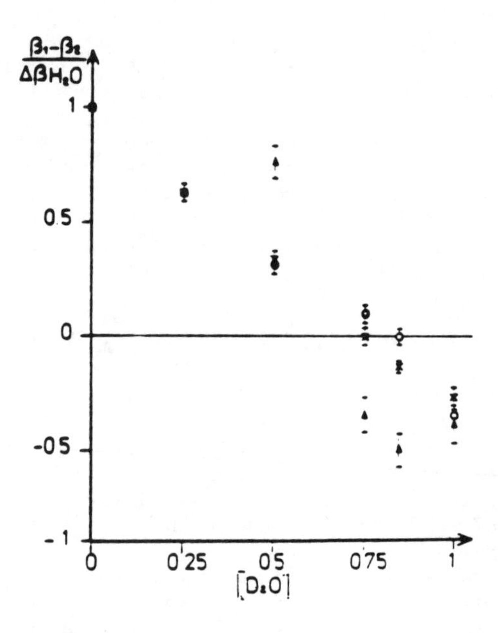

Figure 5. Multiplication used in superposition of Figures 2-4 as a function of $[H_2O]/[D_2O]$.

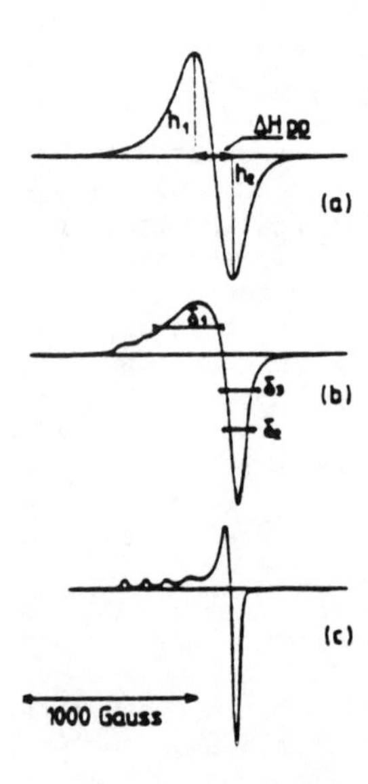

Figure 6. ESR spectra of reference solutions at 77 K with different Cu^{++} concentration (g Cu^{++}/cm^3): a, 0.19; b, 0.105; c, 0.042; and definition of the different parameters used to characterize the spectra. Reproduced with permission from Ref. 4. Copyright 1983 J. Appl. Polym. Sci.

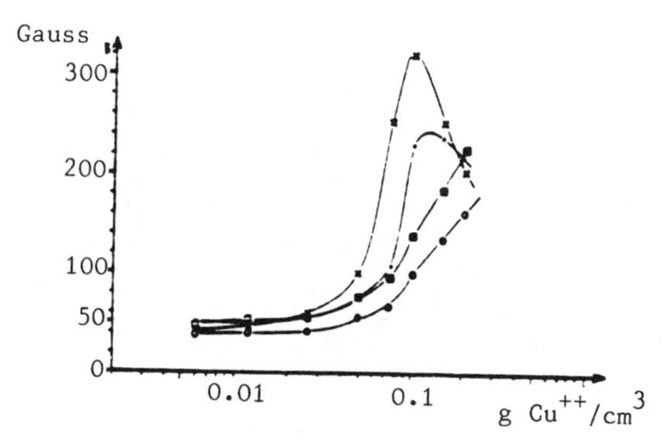

Figure 7. Changes in the width spectral parameters vs. Cu^{++} concentration of the reference solutions. Key: X, δ_1; o, δ_2; □, δ_3; •, Δ, H_{pp}. Reproduced with permission from Ref. 4. Copyright 1983. J. Appl. Polym. Sci.

dipole-dipole and exchange interactions are short range, the measured concentrations are local concentrations on a few tens of Å scale. Use of cryoprotectors permits us to have homogeneous solutions so that the local and average concentrations are the same in the reference solutions. However, in the membranes where we expect ionic segregations, the local concentrations measured by ESR should be different from the average concentration defined from chemical analysis.

In Table I are summarized the results obtained for different sample: Nafion perfluoroethylene membrane (NAF), acrylic acid irradiation grafted polytetrafluoroethylene (PTFE), sulfonated styrene irradiation grafted fluorinated ethylene propylene copolymer (RAI), and sulfonated polysulfone (SPS).

Table I. Average and local Cu^{++} concentrations in membranes corresponding to different degrees of exchange and different water contents

Exchange capacity meq/g	Specimen	Average concentration [Cu] g/cm^3	Water content g/g	Local [Cu^{++}] (g/cm^3)	Concentration ratio
0.8	NAF	0.047	0.045	0.21	4.46
0.8	NAF	0.039	0.15	0.115	2.96
0.8	NAF	0.018	0.03	0.082	4.53
3.33	PTFE	0.049	0.061	0.088	1.8
3.33	PTFE	0.016	0.095	0.065	4.06
3.2	PTFE	0.055	0.047	0.045	2.65
3.2	PTFE	0.01	0.148	0.048	5.03
1.2	RAI	0.0524	0.03	0.208	3.97
1.2	RAI	0.0518	0.03	0.21	4.05
1.2	RAI	0.0652	0.03	0.21	3.22
1.2	RAI	0.0422	0.15	0.094	2.84
1.2	RAI	0.0417	0.15	0.0987	2.36
1.2	RAI	0.0525	0.15	0.105	2
1.0	SPS	0.027	0.015	0.096	3.55
1.0	SPS	0.022	0.23	0.077	3.5
0.75	SPS	0.0063	0.03	0.042	6.6
0.75	SPS	0.0125	0.03	0.046	3.68
0.99	SPS	0.0227	0.04	0.0924	4.07
0.99	SPS	0.0272	0.04	0.113	4.15

Let us discuss some of the results. The local concentrations are always larger than the average concentration obtained from chemical analysis. This is direct evidence for nonrandom distribution of ions in the membranes. For instance, in the low water content Nafion sample we found a ratio larger than 4, which means that the ionic phase represents less than 25% of the polymer. Such a large ionic concentration cannot be explained only by a phase segregation of the ether comonomer, which would give a factor smaller than 2. Therefore, the ionic phase has to contain Cu^{++} ions, water, and only part of the side chain. Another interesting point is the change in local concentration when the water content is changed. If the local Cu^{++} concentration in the sample containing 4.5% water by weight is 0.21 g/cm^3, we can calculate the new Cu^{++} concentration if we assume that when the water content changes from 4.5% to 15% all the new absorbed water molecules are absorbed in the phase containing Cu^{++} ions. With such an hypothesis we would expect to observe a change in the local concentration from 0.21 g/cm^3 to 0.107 g/cm^3. This number is pretty close to the experimental value of 0.115 g/cm^3 and is consistent with the hypothesis that most of the water molecules are absorbed in the ionic phase.

We also have to compare the local Cu^{++} concentration with the concentration obtained by assuming that we have a ionic phase containing only water and Cu^{++} ions in this hydrated sample. The 15% water content sample contains 20.5 water per Cu^{++} and the corresponding solution would therefore have a concentration of $63(63 + 20.5 \times 18) = 0.145$ g Cu^{++}/g. The concentration in g/cm^3 will be larger because the density is greater than 1. This value is definitely larger than the local concentration found in this sample (0.115 g/cm^3).

The absolute values of the local concentration are given within 20% accuracy. The relative values corresponding to changes in water contents are obtained with an accuracy better than 5%.

All this evidence indicates a phase segregation of the Cu^{++} ions in this membrane. The local Cu^{++} concentration is three times larger than the average concentration in the fully hydrated specimen, and this ratio is still larger ($\sim$ 4.5) for the low water content specimens. The changes observed in the local concentrations when the water content is changed are consistent with the hypothesis that most of the water molecules are going inside this ionic phase. Since the local Cu^{++} concentration has been shown to be smaller than the Cu^{++} concentration obtained from a ionic phase containing only water and Cu^{++} ions, some other organic groups have to be included in this phase.

Similar ratios of the local ion concentration versus the average concentration have been obtained for the other membranes. For some membranes the definition of the local concentration is more difficult, probably because of superposition of different ionic species.

These electron spin resonance studies directly show the existence of phase segregation in these ion exchange membranes. The ionic phase is made of the ions, water molecules, and part of the side chains.

Mössbauer spectroscopy is a powerful technique to give information on the ionic phase after exchange with specific cations like ^{57}Fe, ^{151}Eu or ^{113}Sn. Most of our work on the perfluorinated ionomer membranes has been done with Fe^{++} and Fe^{+++} ions (5,6). A typical resonant absorption experiment uses a monochromatic radioactive Mössbauer source which emits γ rays; the sample to be studied is used as an absorber and must contain nuclei of the same stable isotope emitted by the source. Resonant absorption occurs whenever the difference in energy between the ground and excited states of the nuclei in source and absorber precisely coïncide. Each state is split by hyperfine interaction of the nuclear electric and magnetic moments with the electric and magnetic fields created at the nucleus by its surrounding electrons and more distant atoms. The energy of γ rays emitted by the source is slightly modulated; the spectrum is scanned by varying the Doppler shift obtained by moving the source with a velocity, v, 10 mm/s.

Different information can be extracted from the absorption spectra. The area of the absorption spectrum is governed by the probability of a nucleus absorbing a γ photon emitted by the source without recoil. The recoiless fraction is:

$$f = e^{-E_\gamma^2 \langle x^2 \rangle / h^2 c^2}$$

where $\langle x^2 \rangle$ is the mean square displacement of the absorbing nucleus. Absorption disappears entirely when $\langle x^2 \rangle$ diverges at the melting point or near the glass transition of a noncrystalline phase. The quadrupole splitting values reflect the existence of an electric field gradient produced by the asymmetry in the electrons or neighbouring atoms distribution. Structural information can also be derived from the effects of magnetic interactions on Fe^{+3} spectra. A paramagnetic hyperfine structure can be observed when the distances between Fe^{+++} ions are large enough to give long electronic relaxation times. Small iron rich clusters may order magnetically and lead to superparamagnetism above a certain blocking temperature.

Mössbauer spectroscopy results with Fe^{++} and Fe^{+++} perfluorinated ionomer membranes can be summarized. First of all, Mössbauer absorption disappears at temperatures usually well below 0°C. Such a result already excludes the existence of iron ions dispersed in a perfluorinated matrix for which we would observe a disappearance of absorption at a temperature corresponding to the glass transition of the matrix ($\sim$ 300°C). In Figure 8 are plotted the changes in $\ell n f$ versus temperature for different water contents. A deviation from the straight line corresponding to a Debye model with θ = 140 K appears at a temperature associated with a glass transition of the ionic phase containing water and ions. Such a result is consistent with the NMR results and quasi elastic neutron scattering from which water protons have been found to be mobile down to temperatures of the order of 200 K.

In Figure 9 are shown the low temperature parameters of a ferrous Nafion as a function of water content. A drastic change occurs between 0 and 8% weight water content for all the parameters. This result, associated with Na^+, H^+ NMR results and with heat of sorption changes, implies that the first water molecules are absorbed specifically at the cations.

Through analysis of the microstructure of the ionic phase in the Fe^{+++} exchanged membranes different ionic species have been identified:

- isolated Fe^{+3} ions giving a paramagnetic hyperfine spectrum;
- dimers giving a specific doublet with a large quadrupole splitting;
- iron in groups containing even numbers of antiferromagnetically coupled ferric ions.

Evidence of the existence of a separate ionic hydrophilic phase has therefore been obtained from these different measurements. The local structure inside this ionic phase seems to depend strongly on the nature of the cation. Specific associations have been found for Fe^{+++} ions. The question which arises now concerns the geometry of these ionic associations. This problem will be developed in the next chapter.

Dimensions of the Ionic Domains

A typical behavior observed in ionomers is the existence of a small angle scattering peak (from X ray or neutron experiments) and a zero

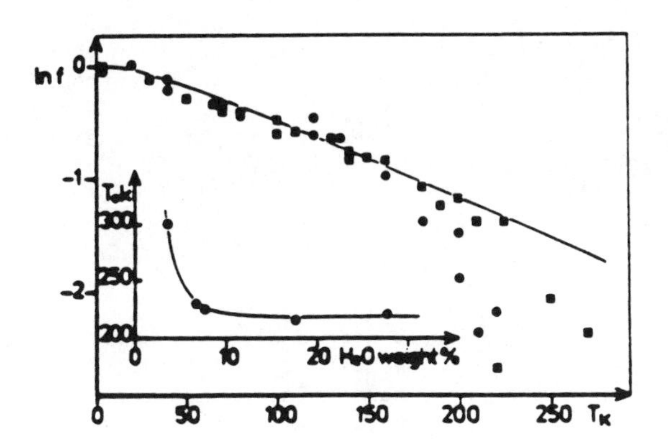

Figure 8. ℓnf vs. T for a ferrous Nafion as a function of water content. Key: ■, 3.6% H_2O; ●, 6.6% H_2O; □, 7.8% H_2O; o, 17.5% H_2O. The solid line corresponds to the theoretical ℓnf curve with θ_D = 140 K. Insert shows the temperature, T_0, at which the f-factor falls to zero.

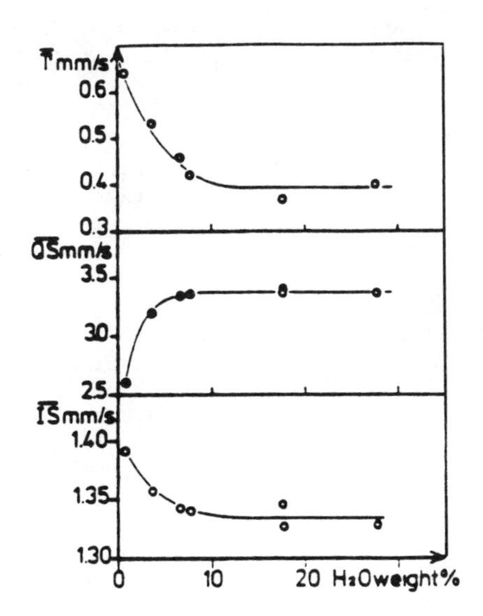

Figure 9. Low temperature (4.2 K) parameters of a ferrous Nafion as a function of water content (one-site fit). Reproduced with permission from Ref. 9. Copyright 1984 J. Reactive Polym.

order scattering increase. Both the intensity and the position of the peak change with water content. This peak has therefore been associated with the ionic hydrophilic domains. Analysis of this peak in terms of a Bragg peak resulting of the presence of a polycrystalline lattice of spherical ionic aggregates has been made by T. Gierke (7) for the Nafion perfluorinated materials. The proposed model also based on ion transport implies the existence of these spherical ionic clusters separated by small channels which permit the ion and water diffusion. In a similar system, Fujimura (1) concluded that the general aspects of the variation of the scattering profiles with swelling and deformation can be qualitatively described by one of the following models: (i) a two phase model in which spherical ionic aggregates are dispersed in the polymeric matrix or (ii) the core shell model.

In a another study (8), we have analyzed the literature data associated with our SANS data and shown that the scattering data on perfluorosulfonated ionomer membranes are consistent with the scattering produced by a group of hard spheres dispersed in the polymeric matrix. The number of ions per cluster was found to change with the water absorption values, with the cation, and with equivalent weight. The occurence of the scattering maxima is due to interference effects between clusters. Further calculations have to be made to take into account a possible anisotropy of these ionic spheres.

Attempts to Label the Ionic Domains

We have completed experiments label the ionic domains. We have found evidence of a precipitation phenomenon of particles of iron oxide or hydroxide when an iron form of membrane was exchanged by different other ions like K^+, Na^+, etc. We therefore have analyzed these particles by different techniques -like X rays, Mössbauer spectroscopy, magnetic measurements and electron microscopy- with two goals in mind. First of all the formation of ultra thin particles is very important in different domains and especially in catalysis when these membranes are used in the solid polymer electrolyte process. Second, we expect some correlation between the sizes and distribution of precipitates with the starting ionic domains.

Let us summarize here some previous results (9,10). When soaking a Fe^{+3} membrane like Nafion 1200 in a solution of K^+, the exchange is effected but the iron is not eliminated from the membrane. We indeed have a Mössbauer absorption but a very different spectrum (Figure 10). All of the iron is now contained in a single new phase which displays a magnetically split hyperfine pattern at 4.2 K. The structure of this new phase depends on different parameters, the most important of which seems to be the initial water content. Amorphous ferric hydroxide has been obtained for low water contents. αFeOOH (goethite) is obtained for high water content (118%). αFe_2O_3 (hematite) is obtained when a sample first exchanged with Fe^{+3} is autoclaved at 150°C.

The distribution of these precipitates across the membrane thickness is not uniform, as has been shown by electron microprobe Figure 11). The profile form depends on different parameters such as time of soaking in KOH solutions, KOH concentrations, etc.

These particles of αFeOOH and αFe_2O_3 have been observed directly

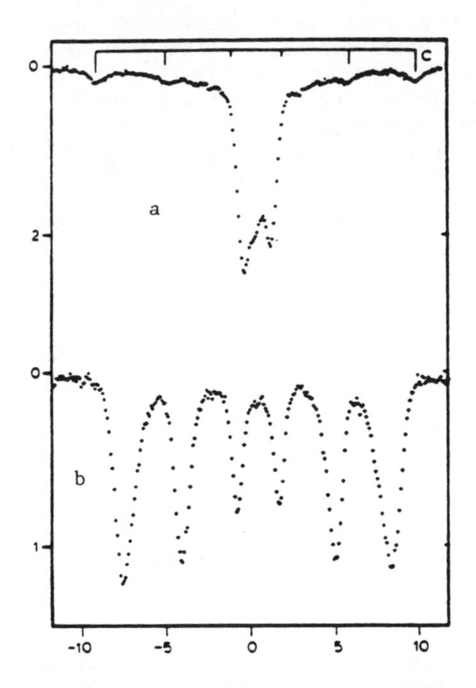

Figure 10. Mössbauer spectra at 4.2 K of an Fe^{+3}. Nafion sample before (a) and after (b) reexchange with K^+. Reproduced with permission from Ref. 9. Copyright 1984 J. Reactive Polym.

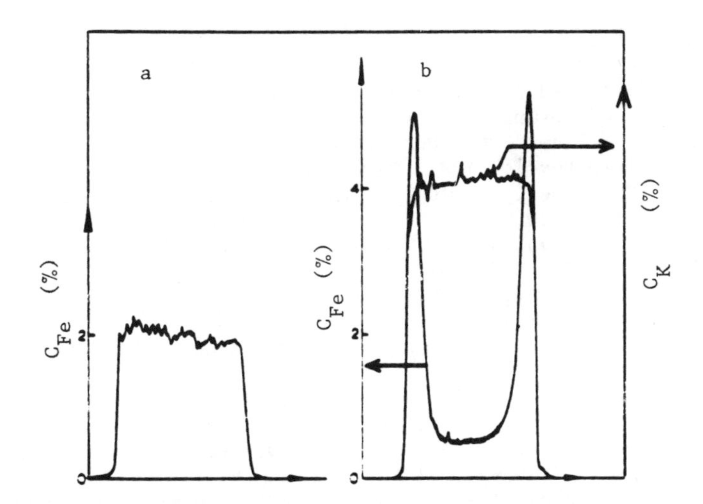

Figure 11. Distribution of iron across the thickness of the membrane before (a) and after (b) exchange with K^+.

in ultra thin (500 Å) sections of these perfluorosulfonate membranes by transmission electron microscopy (TEM). αFe_2O_3 particles (∿ 100 Å in diameter) are roughly spherical; they are grouped together in clusters ranging from a few hundred up to about 1000 Å. These clusters are uniformly distributed across the membrane thickness. The αFeOOH particles are in the form of well separated acircular or blade shaped crystallites up to 1000 Å long and about 100 Å across.

These results may constitute a first step toward a direct observation of ionic domains.

Conclusions

Many organic polymeric ion exchange membranes have a three phase structure:

- microcrystallites;
- intermediate hydrophobic phase;
- hydrophilic ionic domains that constitute the active part of the membrane for the ion exchange process.

We have given many experimental results concerning this ionic phase in terms of chemical composition, dynamic properties, microstructure of the ionic complexes and geometry of the domains.

Acknowledgments

The work described in this article summarizes general results obtained by different coworkers including Drs. J.M.D. Coey, B. Rodmacq, A. Meagher for Mössbauer spectroscopy; Drs. E. Roche, R. Duplessix and S. Kumar for SANS experiments; Drs. F. Volino and D. Galland for ESR; Drs. J. Kelly, A. Michas, J.Cl. Jesior for precipitation studies, Drs. B. Dreyfus and M. Escoubes for general discussions on thermodynamic results.

Literature Cited

1. "Perfluorinated Ionomer Membranes," Eisenberg, A.; Yeager, H.L., Editors, ACS Symposium Series 180, 1982.
2. Roche, E.; Pinéri, M.; Duplessix, R.; Levelut, A.M. J. Polym. Sci. Polym. Physics 1981, 19, 1-11.
3. Roche, E.; Pinéri, M.; Duplessix, R. J. Polym. Sci. Polym. Physics 1982, 20, 481.
4. Vasquez, B.; Avalos, J.; Volino, F.; Pinéri, M. J. Appl. Polym. Sci. 1983, 28, 1093-1103.
5. Rodmacq, B.; Coey, J.M.D.; Pinéri, M. Revue de Physique Appliquée 1980, 15, 1179-1182.
6. Rodmacq, B.; Pinéri, M.; Coey, J.M.D.; Meagher, A. J. Polym. Sci., Polym. Physics 1982, 20, 602-621.
7. Gierke, T.D.; Hsu, W.Y. Ref. 1, 283-310.
8. Kumar, S.; Pinéri, M., submitted to Journal of Polymer Science.
9. Meagher, A.; Rodmacq, B.; Coey, J.M.D.; Pinéri, M. J. Reactive Polym. 1984, 2, 51-59.
10. Pinéri, M.; Coey, J.M.D.; Jesior, J.Cl., submitted to Journal of Reactive Polymers.

RECEIVED November 27, 1985

RUBBERS AND SOLUTIONS

14

Ion-Hopping Kinetics in Three-Arm Star Polyisobutylene-Based Model Ionomers

Masanori Hara[1,3], Adi Eisenberg[1], Robson F. Storey[2,4], and Joseph P. Kennedy[2]

[1]**Department of Chemistry, McGill University, Montreal, Quebec, Canada H3A 2K6**
[2]**Institute of Polymer Science, University of Akron, Akron, OH 44325**

Three-armed ionically terminated "stars" are used to determine the kinetics of ion-hopping or interchange. These species form a network in which the ionic aggregates provide the weak links. Stress relaxation techniques are utilized, and from the relaxation times the kinetic parameters are calculated. For the sodium sulfonate terminal groups, the first order rate constant is given by $k = 7.11 \times 10^9 \exp(-94{,}000/RT)$ with ΔH in Joules/mol.

It has long been recognized that the flow of ionomers, because of the presence of ionic aggregates, is related to the rate at which ionic groups (pairs) remove themselves from one aggregate and move to another (1-2). Since ionomers have recently become of great industrial and academic interest, and since the flow of ionomers represents a fundamental problem in study of mechanical properties as well as the production of these materials, it has been of great interest to determine the rate at which this ion-hopping process occurs. Attempts to study this fundamental process in bulk ionomers have not yielded much quantitative information, because, in most ionomer systems, two different types of aggregates are thought to be present i.e. multiplets and clusters (3-4). Multiplets are relatively small ionic aggregates consisting of a few ion pairs, while clusters can be quite large and presumably also contain considerable organic chain material. Thus, in order to obtain information about ion-hopping kinetics, it is necessary to work on a system in which only multiplets are present. Since conventional ionomers, i.e., polymers containing a large concentration (4-12 mole %) of randomly placed ionic groups, generally contain both clusters and multiplets (5-6), other systems have to be considered.

Difunctionally terminated polymer chains are, obviously, of great interest in this context. Teyssie and coworkers (7-15) have investigated the properties of halato-telechelic systems extensively, including a number of rheological studies. They found that, for

[3]Current address: Rutgers, The State University of New Jersey, Department of Mechanics and Materials Science, Piscataway, NJ 08854
[4]Current address: Department of Polymer Science, University of Southern Mississippi, Hattiesburg, MS 39406–0076

0097–6156/86/0302–0176$06.00/0

polybutadiene halato-telechelic systems with divalent cations (Ba, Ca, Zn, and Mg), only one relaxation mechanism characteristic of ionic aggregates worked without chain flow due to the layered structures of multiplets. The activation energies obtained are from 62.8 KJoul/mole for Ba salt to 128 KJoul/mole for Mg salt. However, in general, it is probable that ionic aggregation in the halato-telechelic systems with monovalent cations consists mainly of ion quartets (involving only two chain ends), as shown schematically in Figure 1. In this case, isolation of the ion-hopping mechanism from additional relaxation mechanisms, e.g., normal chain flow, cannot be assured. It is therefore desirable to study a system in which the polymer is present as a covalently crosslinked network, the chains of which contain "weak" or thermolabile linkages in the form of ion quartets. A three-arm star polymer carrying terminal ionic groups, also shown in Figure 1, can yield such a system provided the ionic groups are present in small enough concentration so that clustering does not occur. Such a system would form a network which satisfies all the requirements for the measurement of ion-hopping kinetics.

Three-arm star ionomers of this type have recently become available (16-17). They consist of polyisobutylene (PIB) chain segments emanating from a central phenyl ring and carrying metal sulfonate groups at the three chain ends. Molecular weights of these polymers range from 6,000 to 20,000. The chemistry of these systems has been described extensively. It has been shown that the materials behave like conventional covalently crosslinked rubbers, and investigations of these materials are continuing (18-23).

A well defined way of obtaining kinetic data for chemical reactions which lead to chain scission in bulk network polymers is by stress relaxation. The formalism of this method has been developed by Tobolsky and others (24-26), and can be summarized as follows.

From the kinetic theory of rubber elasticity, the stress, f(t), at time t is related to the initial stress, f(0), by where N(t) is

$$\frac{f(t)}{f(0)} = \frac{N(t)}{N(0)} \tag{1}$$

the number of network chains per unit volume supporting the stress at time t.

When the network chains are undergoing an interchange reaction, first-order reaction rate law can be applied, i.e.

$$-\frac{dN(t)}{dt} = k\,N(t) \tag{2}$$

where k denotes the rate constant of the interchange reaction. Integration of Equation 2 with the boundary condition N(t)=N(0) at t=0 gives

$$\frac{N(t)}{N(0)} = \exp(-\,kt) = \exp(-t/\tau) \tag{3}$$

where $k=1/\tau$ and τ denotes the relaxation time. Combining Equation 1 and Equation 3 yields the law for Maxwellian decay.

Finally, the rate constant k for the interchange reaction can be expressed by the following equation

$$k = A \exp(-\Delta E_a/RT) \tag{4}$$

where ΔE_a is the activation energy of the reaction and R is the gas constant.

The assumptions made in applying these equations to ion-hopping reactions are that all the network chains are of uniform length and that the ion-hopping mechanism is reflected only in the "ultimate equivalent" Maxwell element.

Experimental

Three-arm star PIB ionomers were prepared by techniques described extensively before (16-17). Some relevant parameters for the sample are shown in Table I. The 3-arm PIB ionomers neutralized by NaOH were dried at room temperature under vacuum for 1 month, while another series of Na salts were dried at 140°C for a week. Samples were then compression molded at 150°C under an applied load of 4000 lbs/in^2. Typical dimensions of the rectangular specimens for stress relaxation studies were 3.0 x 6.0 x 40 (mm).

The experiments were performed on a stress relaxometer (27) under a nitrogen atmosphere. The temperature inside the sample chamber was constant within ± 0.2°C in the range of 60 to 150°C. The stretching mode was utilized and the elongation was ca. 10% of total length.

Previous studies of ionomers have shown that dynamic mechanical tests are most sensitive to the presence of large clusters. For this reason, dynamic mechanical test was performed using a computerized torsion pendulum (28). The frequencies varied from ca. 3 Hz for the glassy region to ca. 0.1 Hz for the low modulus region. The heating rate was usually 0.6°C/min with a temperature control of ± 1°C.

Results

Figure 2 shows the shear storage modulus, G', the shear loss modulus, G", and tan δ for the sample dried at room temperature. The primary transition is observed around -55°C and a well-defined rubbery plateau is observed from -20 to 140°C. The results of the stress relaxation experiments at various temperatures, plotted as log $E_r(t)$ vs. t, are shown in Figure 3. It is seen that the terminal portions of all the curves are indeed linear; thus, the maximum relaxation time at each temperature, τ_m, can be obtained from the slope of the curves. The maximum relaxation times are plotted as log τ_m vs. 1/T in Figure 4. Linearity is observed within experimental error over three orders of magnitude of time. The activation energy estimated from the slope of the curve is 94 KJoul/mole. Figure 4 also shows the data for samples dried at high temperature. It is seen that the slopes of the two sets of data are identical. The position of the plot for the sample dried at high temperature is, however, much higher than for that dried at a lower temperature. These plots are very reminiscent of classical rheological data from which kinetic parameters of the ion interchange process can be calculated (29-33). The kinetic data obtained for the ion-hopping reaction of 3-arm star PIB ionomer is $k = 7.11 \times 10^9 \exp(-94{,}100/RT)$ for the sample dried at room temperature and $k = 6.90 \times 10^8 \exp(-94{,}100/RT)$ for the sample dried at high temperatures.

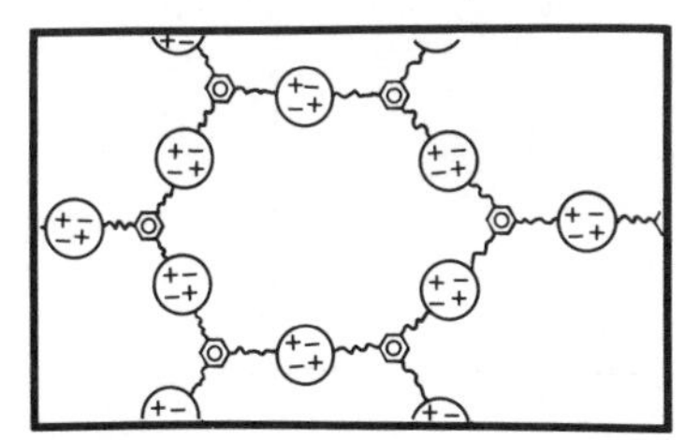

Figure 1. Schematic diagrams comparing halato-telechelic (linear) ionomers vs. three-arm PIB ionomers.

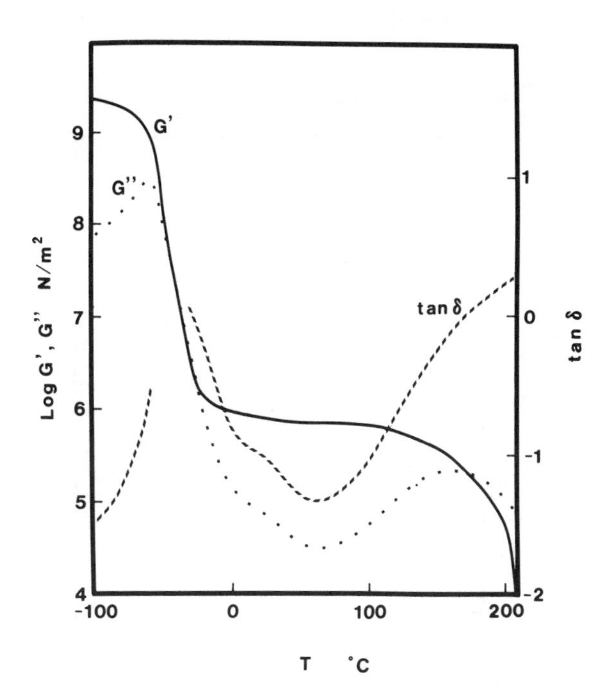

Figure 2. Variation of the shear storage modulus, G', the shear loss modulus, G", and tan δ for the sample dried at room temperature.

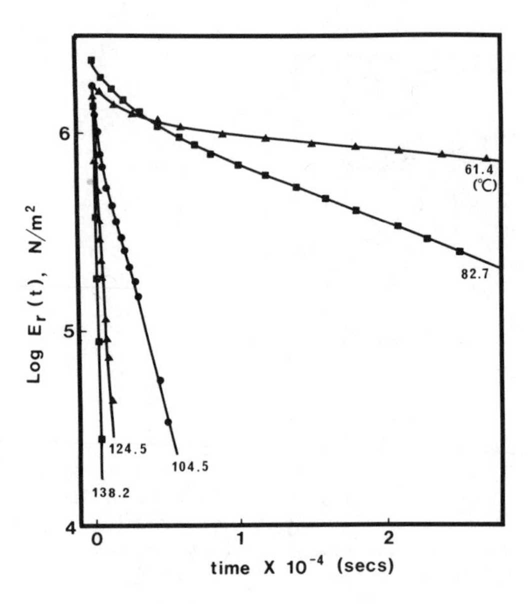

Figure 3. Relaxation modulus (Er(t)) vs. time for the sample dried at room temperature.

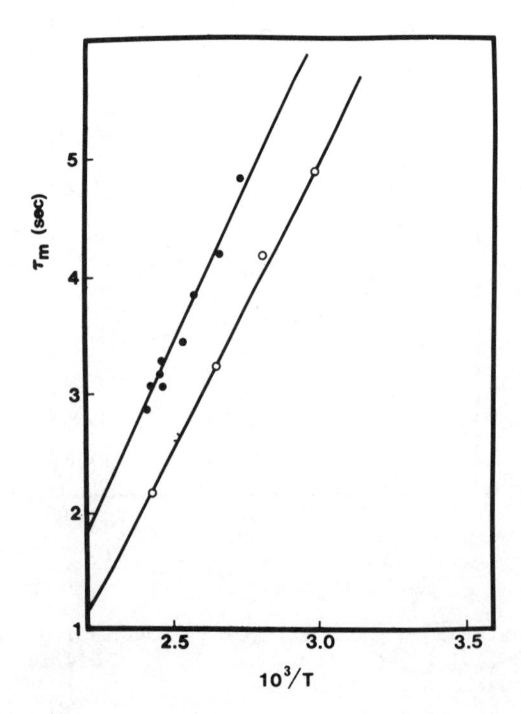

Figure 4. Maximum relaxation time (τm) vs. 1000/T for samples dried at room temperature (O) and at 140°C (●).

Table I. Relevant parameters for 3-arm PIB star ionomer.

Parameter	Symbol	Value	Method
Molecular Weight	$(\bar{M}n)$	6,900	VPO
Functionality	$(\bar{F}n)$	3.2	titration
Glass Transition Temperature	(Tg)	-56°C	DSC
		-55°C	T.P.

Table II. Temperature dependence of Em and Mc for samples dried at room temperature.

Temperature (°C)	Em (N/m^2)	Mc^{exp} (a)	Mc^{exp}/Mc^{cal} (b)
61.4	1.05×10^6	6,670	1.45
82.7	1.34	5,560	1.21
104.5	9.0×10^5	8,780	1.91
124.5	7.3	11,400	2.48
138.2	7.4	11,600	2.52

(a). Mc^{exp} was obtained from the equation Em = 3ρRT/Mc.
(b). Mc^{cal} was calculated to be 4,600 (= 6,900 x 2/3), assuming that all ions are present as quartets.

Table III. Temperature dependence of Em and Mc for samples dried at 140°C.

Temperature (°C)	Em (N/m^2)	Mc^{exp}	Mc^{exp}/Mc^{cal}
93.3	4.95×10^5	15,500	3.37
102.7	4.2	18,700	4.07
116.1	4.4	18,500	4.02
120.4	4.3	19,200	4.17
132.3	4.5	18,800	4.09
133.0	3.5	24,300	5.28
134.4	3.5	24,400	5.30
136.1	2.2	39,000	8.48

Discussion

In the 3-arm star PIB ionomers, most of the ionic aggregates should be multiplets due to the low ion content (2.4 mole %) (3-4). Figure 2 shows only the shoulder for secondary transition in the tan δ curve (around 30°C); thus, only very small amounts of ions exist as clusters (35), if they are present at all. Most of the ions are thus expected to exist as small multiplets, i.e. quartets. Moreover, it is difficult for the molecules to flow due to network crosslinks resulting from the presence of the trifunctional linkages. As is seen in the G' curve in Figure 2, the well-defined rubbery plateau extends over a wide temperature range, which is reminiscent of a typical covalently cross-linked rubber (36-37). In the present system, the modulus corresponding to the ultimate Maxweel element is as large as 10^6(N/m^2), which is consistent with the existence of crosslinks rather than simply the entanglement of molecules.

Table II shows the values of Em (the modulus corresponding to the ultimate Maxwell element), and Mc (the molecular weight of the network chain) for the samples dried at a low temperature. It is clear from the table that Mc^{exp}/Mc^{cal} is larger than 1, which means that some of the 3-arm PIB ionomers function as difunctional polymers, i.e. some of the ions are present as free pairs. It is also seen that the higher the temperature, the more of the polymer chains act as difunctional units. This result confirms that ionic aggregates are disrupted by increasing the temperature, a phenomenon fundamental to the use of these materials as thermoplastic elastomers. Moreover, this result parallels the finding of Neppel et al. (5-6) who showed that clusters decompose progressively to multiplets as the temperature is increased.

It was observed that drying the sample at high temperature increased the m and decreased Em, while ΔE_a remained almost the same. The decrease in Em suggests that annealing at high temperature may have led to some chemical degradation of the sample which is further confirmed by Table III when one considers the values of Mc^{cal}/Mc^{exp}. For this reason, the samples dried at room temperature are considered most reliable.

In summary, it is shown that the 3-arm star PIB ionomer is useful in the study of ion-hopping kinetics. The polymer chains are crosslinked by covalent trifunctional linkages and ionic aggregate exist mostly as multiplets in the middle of the chains. The kinetic constants for ion-hopping in this system are given by $k = 7.11 \times 10^9 \exp(-94{,}100/RT)$.

Literature Cited

1. Sakamoto, K.; MacKnight, W.J.; and Porter, R.S. J. Polym. Sci. 1970, Part A-2 8, 277.
2. Eisenberg, A. and Navratil, M. Macromolecules 1973, 6, 604.
3. Eisenberg, A. Macromolecules 1970, 3, 147.
4. Eisenberg, A.; and King, M. "Ion-Containing Polymers"; Academic Press: New York 1977.
5. Neppel, A.; Butler, I.S.; and Eisenberg, A. Can. J. Chem. 1979, 57, 2518.
6. Neppel, A.; Butler, I.S.; and Eisenberg, A. Macromolecules 1979, 12, 948.
7. Broze, G.; Jerome, R.; and Teyssie, P. Macromolecules 1981, 14, 224.
8. Broze, G.; Jerome, R.; and Teyssie, P. Macromolecules 1982, 15, 920.
9. Broze, G.; Jerome, R.; Teyssie, P.; and Marco, C. Polym. Bull. 1981, 4, 241.
10. Broze, G.; Jerome, R.; Teyssie, P.; and Gallot, B. J. Polym. Sci., Polym. Lett. Ed. 1981, 19, 415.
11. Broze, G.; Jerome, R.; and Teyssie, P. Macromolecules 1982, 15, 1300.
12. Broze, G.; Jerome, R.; and Teyssie, P. J. Polym. Sci., Polym. Lett. Ed. 1983, 21, 237.

13. Broze, G.; Jerome, R.; Teyssie, P.; and Marco, C. Macromolecules 1983, 16, 996.
14. Broze, G.; Jerome, R.; and Teyssie, P. J. Polym. Sci. Polym. Ed. 1983, 21, 2205.
15. Broze, G.; Jerome, R.; Teyssie, P.; and Marco, C. Macromolecules 1983, 16, 1771.
16. Kennedy, J.P.; Ross, L.R.; Lackey, J.E.; and Nuyken, O. Polym. Bull. 1981, 4, 67.
17. Kennedy, J.P.; and Storey, F. Am. Chem. Soc. Div. Org. Coat. Appl. Polym. Sci. 1982, 46, 182.
18. Kennedy, J.P.; Storey, R.; Mahajer, Y.; and Wilkes, G.L. Proc. IUPAC, Macro 82 1982 905.
19. Mohajer, Y.; Tyagi, D.; Wilkes, G.L.; Storey, R.; and Kennedy, J.P. ibid 1982, 906.
20. Mohajer, Y.; Tyagi, D.; Wilkes, G.L.; Storey, R.F.; and Kennedy, J.P. Polym. Bull. 1982, 8, 47.
21. Bagrodia, S.; Wilkes, G.L.; Mohajer, Y.; Storey, R.F.; and Kennedy, J.P. Polym. Bull. 1982, 8 281.
22. Bagrodia, S.; Mohajer, Y.; Wilkes, G.L.; Storey, R.F.; and Kennedy, J.P. Polym. Bull. 1983, 9, 174.
23. Tobolsky, A.V.; "Properties and Structures of Polymers"; John Wiley & Sons Inc.: New York, 1960.
24. Tobolsky, A.V.; J. Polym. Sci. 1964, 2 637.
25. Yu, H.; J. Polym. Sci. Polym. Lett. 1964, 2, 631.
26. Mahajer, Y.; Bagrodia, S.; Wilkes, G.L.; Storey, R.; and Kennedy, J.P. Polym. Bull.
27. Eisenberg, A.; and Sasada, T. J. Polym. Sci. 1968, C-16, 3473.
28. Cayrol, B. Ph.D. Thesis, McGill University, Montreal, 1972.
29. Stern M.D.; and Tobolsky, A.V. J. Chem. Phys. 1946, 14, 93.
30. Tobolsky, A.V.; Beevers, R.B.; and Owen, G.D. J. Colloid Sci. 1963, 18, 359.
31. Eisenberg, A.; and Teter, L.A. J. Phys. Chem. 1967, 71, 2332.
32. Eisenberg, A.; Saito, S.; and Teter, L.A. J. Polym. Sci. 1966, C-14, 323.
33. Eisenberg, A.; and Saito, S. J. Macromol. Sci. 1968, A2, 799.
34. Rigdahl, M.; and Eisenberg, A. J. Polym. Sci. Polym. Phys. Ed. 1981, 19, 1641.
35. For example, in the case of sulfonated styrene ionomers (34), the peak height of primary transition is much higher than that of secondary transition below the critical concentration, while the hight of secondary transition peak exceeds that of primary transition peak above the critical concentration.
36. Tobolsky, A.V.; Lyons, P.F.; and Hata, N. Macromolecules 1968, 1, 515.
37. Agarwal, P.K.; Makowski, H.S.; and Lundberg, R.D. Macromolecules 1980, 13, 1679.

RECEIVED January 24, 1986

15

Modification of Ionic Associations by Crystalline Polar Additives

I. Duvdevani[1], R. D. Lundberg[2], C. Wood-Cordova[3], and G. L. Wilkes[4]

[1] Corporate Research Science Laboratories, Exxon Research and Engineering Company, Annandale, NJ 08801
[2] Exxon Chemical Company, Linden, NJ 07036
[3] Allied Corporation, Hopewell, VA 23860
[4] Virginia Polytechnic Institute and State University, Blacksburg, VA 24061

Ion containing polymers such as sulfonated ethylene-propylene-diene (EPDM) polymers which strongly associate in the bulk, require ionic-plasticization for melt processing. It was found that a crystalline additive such as zinc stearate can strongly affect material properties in addition to being a highly effective ionic-plasticizer. Zinc stearate is compatible with sulfo-EPDM even at high loadings (over 30% by weight) and it enhances physical associations as reflected in mechanical properties and swelling characteristics. The morphological structure of zinc-stearate/sulfo-EPDM blends was investigated and zinc stearate was found to microphase separate into small crystallites (less than about 5000 angstroms) which act as a reinforcing filler.

Ion-containing polymers can develop strong attractive networks even at low ionic content (1-4). Polymers which contain sulfonate groups neutralized by metal counterions were shown to form particularly strong networks (5). Considerable attention was given to sulfonated Ethylene-Propylene-Diene (Sulfo-EPDM or S-EPDM) polymers, due to their elastomeric nature (6). The incorporation of sulfonate groups neutralized by metal counterions in EPDM can render materials which approximate a cross linked EPDM at lower temperature and can be melt processable at high temperature. This approach to a so called thermoplastic elastomer is different from previous approaches using block copolymers having hard and soft segment such as in polyurethanes (7), styrene-butadiene block copolymers, (8) or polyester block copolymers (9).

In order to enable melt processing of ion containing polymers, such as S-EPDM, it is necessary to introduce a mechanism that weakens the ionic interactions. This can be achieved by the addition of a polar ingredient that would "plasticize" ionic domains at elevated temperatures only. A variety of such ionic-plasticizers were described by Makowski and Lundberg (10). A particularly attractive combination was found to be zinc stearate with a zinc salt of S-EPDM. It was shown that for such a combination melt

0097-6156/86/0302-0184$06.00/0

rheology was indeed suitable for melt processing and in addition mechanical strength was highly enhanced (10,12). This also enabled further dilution or highly extended compounds of S-EPDM with other ingredients (such as oils, fillers, and other polymers) to obtain thermoplastic elastomers of various properties. Many such compounds are described in the patent literature (11) and an investigation of S-EPDM/Polypropylene blends was reported by Duvdevani, Agarwal and Lundberg (12).

This paper attempts to further explore the modification of ionic associations by a crystalline ionic plasticizer, such as zinc stearate, at the solid state. Mechanical properties, swelling behavior, and morphological aspects were studied in order to better understand the role of such crystalline polar additives.

Experimental

Materials and Preparation

A zinc salt of sulfonated EPDM (Zn-S-EPDM) was prepared in our lab according to methods discussed in earlier publications (6,13). The Zn-S-EPDM used in the present study was designated TP-303 and it had a backbone containing about 55 wt% ethylene, 40 wt% propylene, and 5 wt% diene (ENB) with a weight average molecular weight of about 90,000 and a M_w/M_n ratio in the range of 2-2.5. It was sulfonated to a level of about 30 milli-equivalents per 100g (about 1 mole %). A general description of Zn-S-EPDM is depicted in Figure 1. The zinc ions may be balanced by a combination of sulfonate and acetate groups since a 100% molar excess of zinc acetate was used for neutralization in order to assure complete neutralization.

A technical grade of zinc stearate was used for preparing blends of the Zn-S-EPDM with zinc stearate. Such technical grades contain about 90% stearate. The mixing was done both by internal mixers (Brabender or Banbury) and by mill mixing at about 150-200°C, yielding essentialy identical results as explained below. The zinc stearate level (based on the technical material) was varied from 0 to 50g of zinc stearate per 100g S-EPDM (or 0-33.3 wt%). Materials were designated 303-0 to 303-50 indicating TP-303 (Zn-S-EPDM) and the level of zinc stearate in parts per 100 parts of TP-303. Samples were prepared by compression molding at 150°C.

Mixing zinc stearate with Zn-S-EPDM is not as critical as blending two high molecular weight ingredients. This is due to the solubility of molten zinc stearate in polyolefins such as EPDM. It should be noted that when zinc stearate is melt mixed into EPDM, compatibility is evident during the mixing process. However, after cooling to ambient temperatures zinc stearate tends to bloom out of the blend and settle on the surface. On the other hand, there was no such blooming out of the Zn-S-EPDM blends even at the highest loading of zinc stearate. This aspect was also part of the study presented here.

Measurements

Mechanical properties (stress-strain) were measured on an Instron tester at a cross head speed of 10 mm/min. Dumbbells shaped samples were cut with a die and had an 8 mm test length and a thickness of no more than 0.5mm. Additional details may be found in the thesis work of one of the coauthors (C. W. C.) (14).

Swelling measurements were done in heptane. Disc shaped samples of about 0.5 mm in thickness which were immersed in heptane were weighed periodically after wiping all liquid from the sample surface. Results were converted to volume swell in percent.

Scanning-transmission electron microscope measurements were done on microtomed or razor cut samples, without staining.

Differential scanning calorimetry measurements were done in a Perkin-Elmer DSC-2 calorimeter at a scanning rate of 20°C per minute.

Results and Discussion

The introduction to this paper outlined the need for a polar additive to S-EPDM for melt processing. It should be self evident that if such an additive prevents good network formation at the use temperature, the material properties could deteriorate to those of the backbone alone (with no ionic groups). This would be equivalent to a reduction in sulfonate content as was shown by Agarwal, Makowski and Lundberg (15). One needs, therefore, to find a polar additive that could destroy or weaken the network at melt processing temperatures but would enable re-establishment of networks at use temperatures. A convenient way of doing just that is the use of a polar material which is highly crystalline as are some of the stearates and particularly zinc stearate. Such a polar additive is indeed a highly effective agent for melt viscosity reduction as shown by the various workers mentioned in the introduction. On the other hand, the strong tendency of a highly crystalline material as zinc stearate to crystallize below its melting point, helps to remove most of the additive from shielding the ionic groups which can then re-establish the network. The added advantages, as mentioned above, are the non migratory behavior of zinc stearate in Zn-S-EPDM and the property enhancements. This is what will be discussed here in view of the recent measurements.

It was previously shown by Wagener and Duvdevani (16) that zinc stearate incoporated into Zn-S-EPDM exists in the blend in a highly crystallized form. Figure 2 shows that the wide angle x-ray pattern in a fiber which is melt spun out of a 303-50 blend is essentially that of the zinc stearate powder pattern. Figure 3 shows that the DSC thermogram of 303-50 is also very close to that of zinc stearate while 303-0 shows no endothermic peaks. The x-ray studies as well as DSC studies shown in Figure 4 indicate that the crystalline structures are purer or larger as the zinc stearate loading in the blend increases. This can be seen from melting point shifts and peak width in the DSC thermograms or the sharpness of the x-ray ring patterns. It was also shown by Wagener and Duvdevani that zinc stearate can orient in a strecthed fiber and that the orientation relaxes slowly under load (Figure 5).

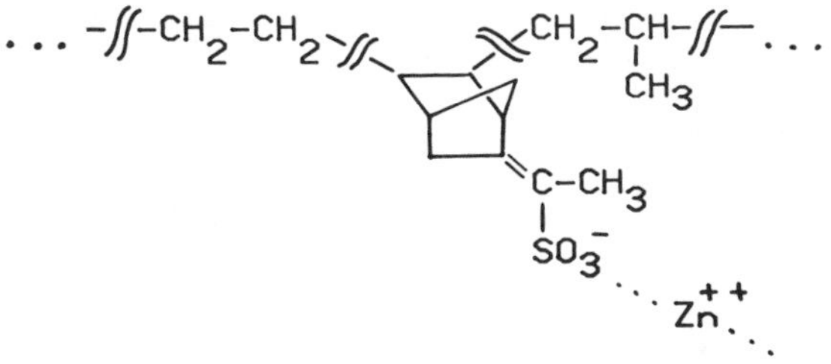

Figure 1. General representation of a Zinc-Sulfo-EPDM molecule (Zn-S-EPDM).

Figure 2. Wide Angle X-ray (WAXS) pattern of 303-50 extruded fiber (right) and of zinc stearate powder (left) (adapted from Ref. 16).

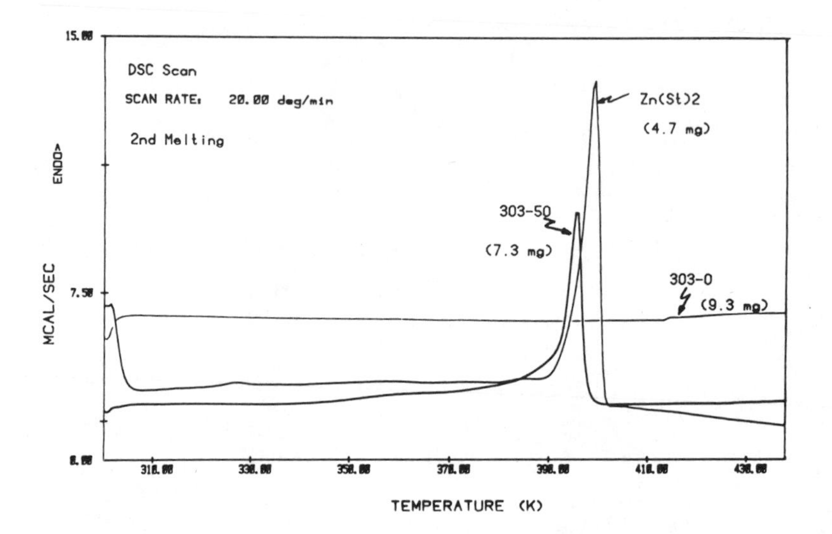

Figure 3. Differential scanning thermograms of 303-0, 303-50 and zinc stearate powder in a second melting pass.

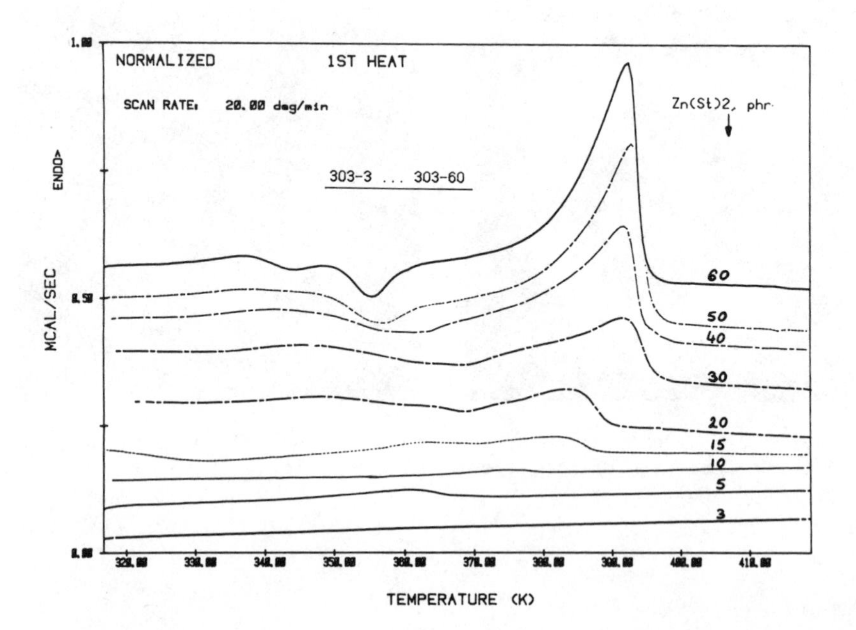

Figure 4. Differential scanning thermograms of Zn-S-EPDM with various loadings of zinc stearate.

Properties

In this investigation we carefully studied the stress strain characteristics as a function of zinc stearate loading and aging. Figure 6 compares stress-strain behavior of 303-0, 303-50 and the vulcanized EPDM backbone. It can be seen that 303-0 is stiffer and stronger than a vulcanized backbone due to the formation of a denser network by ionic association than by sulfur cross-linking. A large number of sites on different chains can be aggregated in an ionic "cross-link" while only two sites are connected in sulfur vulcanization. This point was previously explained by Agarwal and co-workers (15) for the case of S-EPDM. Figure 6 also shows a large increase in modulus obtained by the addition of zinc stearate. Figure 7 shows the stress strain behavior as a function of zinc stearate level for well aged samples while Figure 8 shows the effect of sample aging for no zinc stearate and 50 parts per hundred of zinc stearate. From these curves and from Figure 9 it is clear that modulus changes significantly not only by increasing zinc stearate level but by aging as well. This indicates that zinc stearate acts more than just a filler and that a strong reinforcing mechanism sets up with time. However, since aging effects were more pronounced within the zinc stearate containing systems, the time dependent behavior may also be due to some additional crystallization of the additive. The reinforcement properties of zinc stearate in Zn-S-EPDM are comparable to those of the hard phase in segmented polyurethanes and much higher than simple fillers as shown in Figure 10. The lower four curves in Figure 10, urethane and salt (17), disc spherulites in a rubbery matrix (18) and glass beads in rubber (19) are representing discrete particle fillers dispersed in a matrix. Such behavior can be predicted by various models suggested in the literature like those of Kerner (20) and of Guth (21) and Smallwood (22). However, in the case of well aged samples of zinc stearate in Zn-S-EPDM the behavior is closer to the segmented polyurethane case were the hard phase is chemically linked to the soft portions. It is also close to the reinforcing behavior of small size, surface active fillers such as carbon blacks and fumed silica or to the upper urethane curve in Figure 10 which is a blend with 0.1 micron size filler, (compared to 15 microns for the salt in the curve below it). This behavior is above that predicted by a series model or some of the other models such as mentioned above. It actually comes much closer to the behavior of a continuous hard phase rather than a dispersed one.

It is necessary to clarify that elastic properties beyond the range of low strain or low stress are somewhat deteriorated by the addition of zinc stearate, as is the case with other reinforcing fillers. Since a possibility of "ion-hopping" or "interaction-hopping" can exist in a "physical cross-link," higher stiffness which brings about higher stresses for similar strains can accelerate the "hopping" process. This can be seen in Figures 11 and 12 describing elongation set and stress relaxation as a function of zinc stearate loading. The amount of elongation set at a given strain is higher at higher loadings but so is the stress. However, in spite of a sharp stress relaxation with high zinc stearate loadings, the stress level which the material can sustain at longer time and at a given strain (100% in the case shown in Figure 12) is still

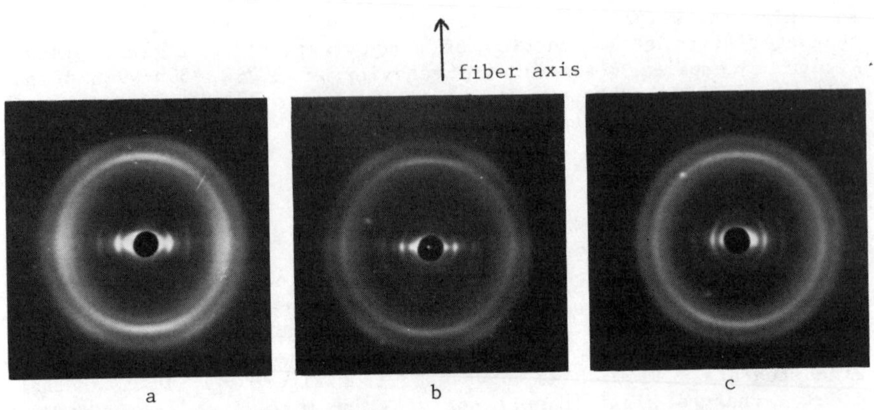

Figure 5. Wide angle x-ray (WAXS) pattern of a 303-50 fiber after it was: (a) stretched to 200% elongation and kept under strain for the WAXS experiment, (b) After 14 days at 200%, (c) unloaded from 200% strain and allowed to relax, (after Wagener et al (16)).

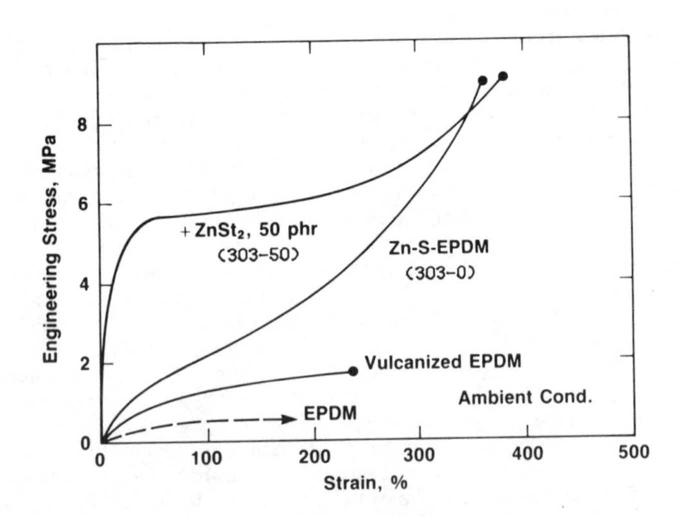

Figure 6. Engineering stress-strain at ambient conditions of 303-0, 303-50 and of the sulfur vulcanized EPDM backbone. ASTM die C samples were stretched at 20 in./min.

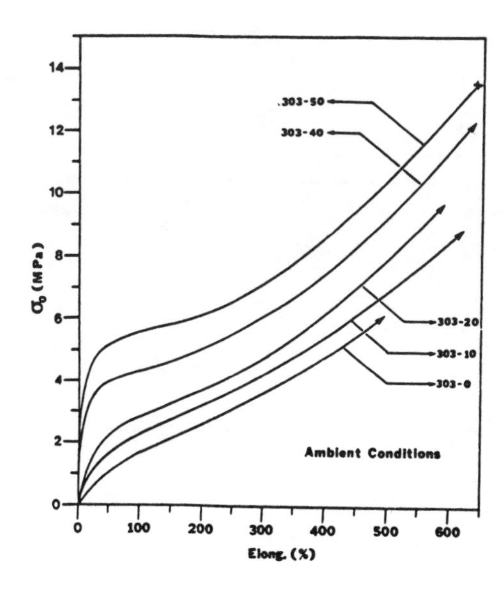

Figure 7. Engineering stress-strain at ambient conditions of Zn-S-EPDM samples with various loading levels of zinc stearate (see text for sample size).

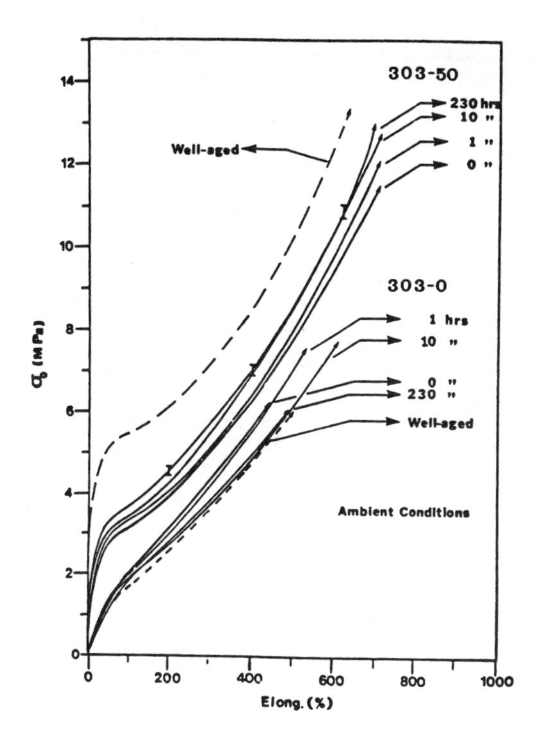

Figure 8. Engineering stress-stain at ambient condition of 303-0 and 303-50 after various aging times.

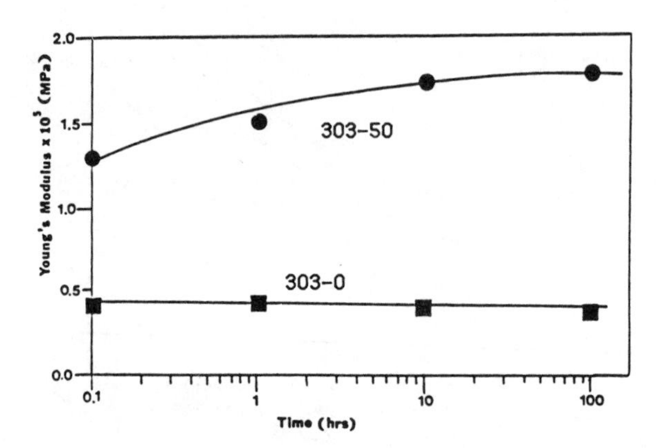

Figure 9. Young's modulus of 303-0 and 303-50 vs. aging time at ambient conditions.

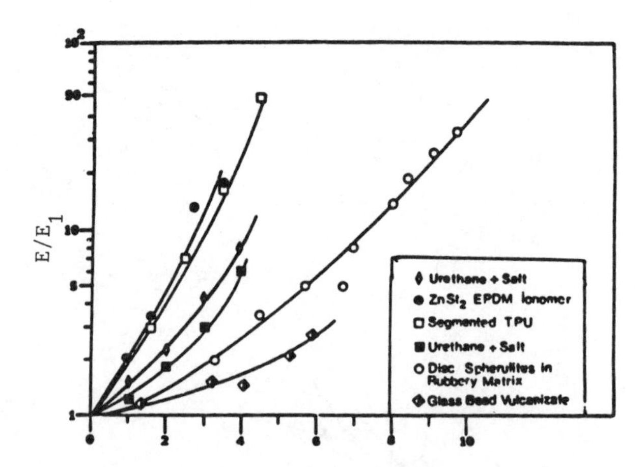

Figure 10. Normalized modulus of Zn-S-EPDM filled with zinc stearate compared to other filled elastomers and segmented polyuretane.

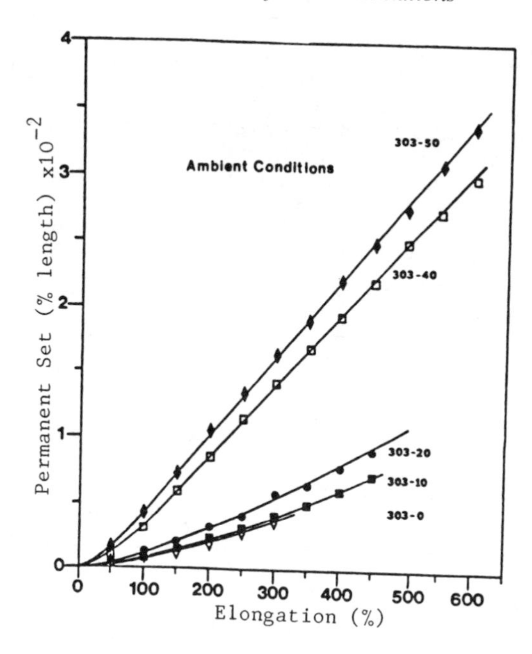

Figure 11. Elongation set of Zn-S-EPDM with various loading levels of zinc stearate as a function of strain level at ambient conditions.

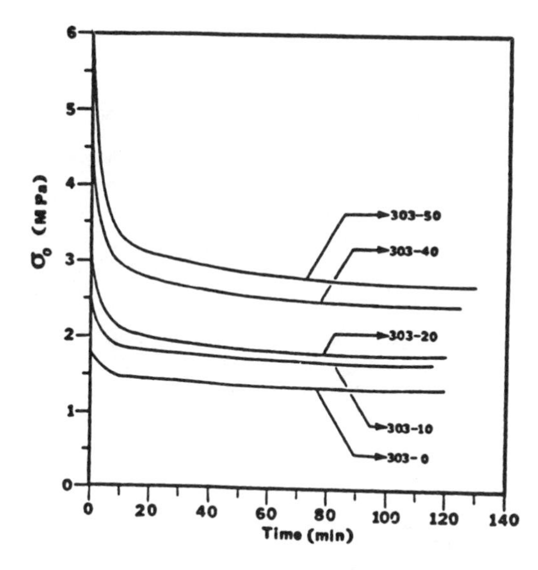

Figure 12. Stress relaxation of Zn-S-EPDM with various loading levels of zinc stearate at a strain level of 100% and at ambient conditions.

higher with higher loadings. This, therefore, is also indicative of a strong interaction between zinc stearate and Zn-S-EPDM even under stress as was found in the x-ray studies of Wagener mentioned above.

An additional measure of the highly reinforcing properties that zinc stearate develops in Zn-S-EPDM was found by swelling behavior. Figure 13 shows volume swell vs time of aged samples containing none, 20 and 50 parts per hundred of zinc stearate in heptane. The slow process of swelling seen in these curves is not due to slow solvent diffusion since thin samples of vulcanized EPDM swelled to equilibrium in heptane within 1 hour. The slow swelling indicates a very slow "ion-hopping" process under the low stresses developed by osmotic pressure.

The swelling data is summarized in Table 1:

Table 1. Percent Volume Swell in Heptane at Ambient Conditions

Material	At 45 min.	At Equilibrium (over 6,000 hrs.)
303-0	160	960
303-20	110	770
303-50	66[2], 90[1]	370
Vulcanized Backbone	180-200	180-200

[1]Freshly molded samples. [2]Aged samples.

This data confirms the strong reinforcing properties imparted by zinc stearate, such that at short swelling times 303-50 had a much denser apparent network then the vulcanized backbone. At equilibrium, 305-50 seems to have a much denser network than 303-0 or 303-20. This is so even when the data is corrected for the volume of zinc stearate which is not swellable in heptane. When this correction is applied the volume percent based on rubber for 303-50 is about 500% at equilibrium which is still significantly less than that of 303-0. Also, it is interesting to note that when 303-0 was blended with fumed silica the swelling results were similar to those of the zinc stearate filled Zn-S-EPDM. Fumed silica is indeed a strong reinforcing filler but it also raises melt viscosity of Zn-S-EPDM significantly in contrast to zinc stearate.

There is a slight difference between freshly molded samples and aged samples in swelling behavior. The freshly molded samples approached equilibrium somewhat faster but it was not noted in very short time measurements, except for 303-50 as indicated in Table 1. It is likely, therefore, that interactions between zinc stearate and sulfonate moieties develop rather rapidly after quenching the melt but a high loading of zinc stearate which makes a large difference in the behavior somewhat prolongs the initial process.

Morphology

In addition to the identification of the crystalline nature of zinc stearate in Zn-S-EPDM as discussed above, we attempted to identify the size and location of these crystals in the polymer matrix. Zinc stearate powder particles tend to be fairly large as can be seen from the SEM micrograph shown in Figure 14. These particles seem to be thin flakes with an overall size of a few micrometers. It is obvious that the crystals in the Zn-S-EPDM matrix are not as large since the materials are fairly clear even at the 50 parts per hundred loading. On the other hand, it is expected that the crystals should be on the order of at least 200 angstroms long since there is a clear ring pattern in the blend (see Figure 2 above) which requires a few unit cell repeats. The unit cell of a zinc stearate crystal from the powder pattern corresponds to dimensions of approximately 8.3 x 5.9 x 45 angstroms. Therefore, we attempted a few electron microscopy techniques to identify these crystals.

An SEM micrograph of 303-50 fractured in liquid nitrogen is shown in Figure 15. Here we could identify some craters which are about 0.1 micrometers in size. These can be suspected as possible sites of zinc stearate crystals but there is no clear evidence for this. The most fruitful technique proved to be that of Scanning/Transmission Electron Microscopy with x-ray analysis. Such STEM micrographs are shown in Figures 16 and 17. When a sample of 303-0, which has no zinc stearate is inspected as in Figure 16, one may recognize 4 typical areas marked in the figure as A-D. Some of the large particles here are similar to what was previously found by Agarwal and Prestridge (23) and by Handlin (24) using TEM micrography. From inspecting peak hight ratios of sulfur, zinc and calcium in the x-ray spectra obtained for the STEM of Figure 16, it can be concluded that area A is different from B, C and D. Area A has a very large ratio of sulfur and calcium and seems to be an inhomogenious area since the rest of the area represented by B, C and D have similar ratios apparently representative by zinc sulfonate with various density of such groups. Area A could be due to calcium sulfate particles that can only be found in the sample infrequently. Figure 17 is a STEM micrograph of 303-50 where the major dark areas are rich in zinc indicating zinc stearate areas. The very few specs of very dark areas are rich in zinc and sulfur and could be zinc sulfate. From Figure 17, it seems that zinc stearate particles are highly distributed in size ranging between a few hundred and a few thousand angstroms. This size range is large enough for obtaining the x-ray pattern which was observed and small enough for causing some turbidity.

It should be noted that Handlin (24) inspected microtomed samples of a similar Zn-S-EPDM modified with zinc stearate, by TEM. The micrographs obtained by Handlin contain dark twisty streaks which he attributed to zinc stearate crystals. He concluded therefore that the zinc stearate appeared to form thin flakes of lamellar crystals about 50 angstroms thick with large lateral dimensions of up to several micrometers. Such a morphology would provide a very large surface to volume ratio with many possible

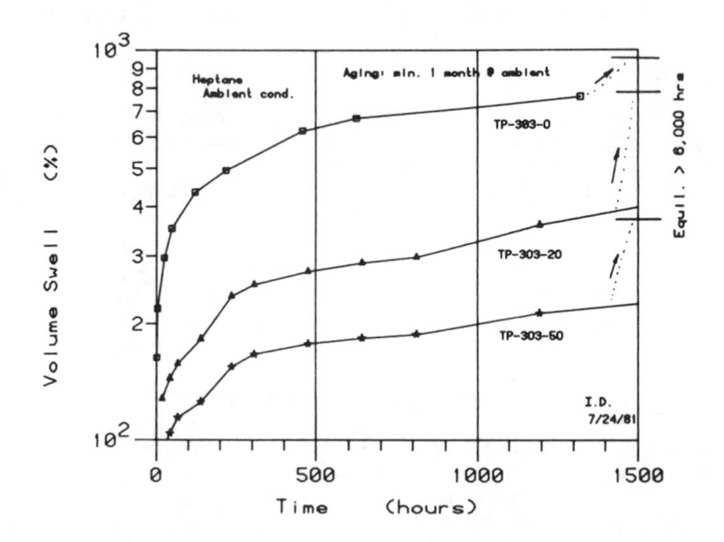

Figure 13. Volume swell in percent for aged samples of 303-0, 303-20 and 303-50 in heptane at ambient condition as a function of time. Values which appeared to level off after 6,000 hours are indicated at right.

Figure 14. Scanning electron micrograph (SEM) of zinc stearate powder (technical).

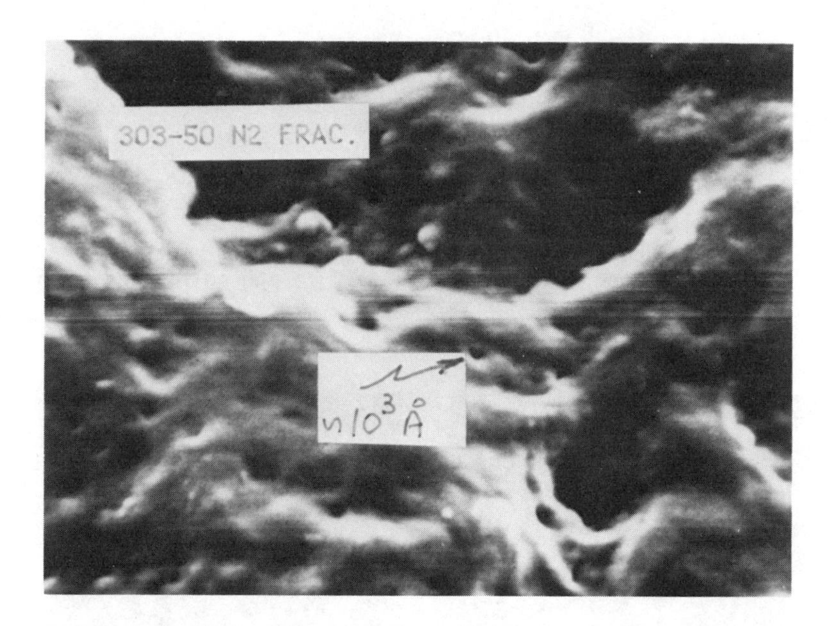

Figure 15. Scanning electron micrograph (SEM) of a 303-50 surface obtained after fracturing in liquid nitrogen.

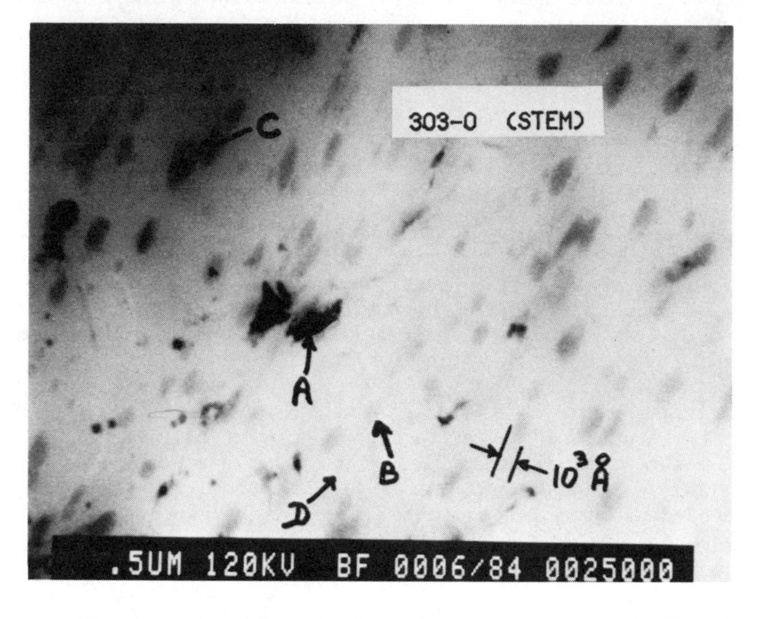

Figure 16. Scanning-transmission electron micrograph (STEM) of a thin film of 303-0 with no staining.

sites for interactions between the zinc stearate and the ion containing regions of the polymer as is suggested below.

Conclusions

From the various data reviewed and presented here, it is evident that a crystalline polar additive such as zinc stearate acts as

- A good "ionic plasticizer" in the melt
- A strong "reinforcing filler" in the solid state, particularly after aging.

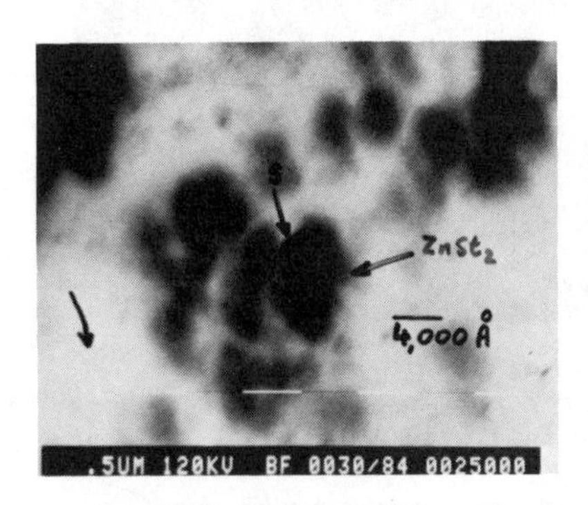

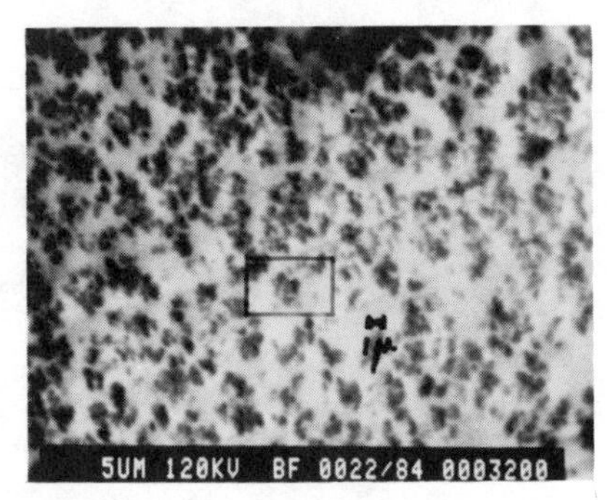

Figure 17. Scanning-transmission electron micrograph (STEM) of a thin film of 303-50 with no staining. The magnified micrograph corresponds to the area indicated on the lower magnification micrograph on the right.

It was also shown that stress relaxes faster with high loadings of zinc stearate but zinc stearate loadings do help to support higher stress levels after long relaxation periods.

These mechanical data along with the x-ray and electron microscopy data could suggest strong interaction between small zinc stearate crystals on preferred crystal planes. One such model is shown in Figure 18. From the x-ray data the long unit cell dimension of zinc stearate is placed at 90° to the stress axis. Stress relaxation can take place through "interaction hopping" but the overall existance of interaction with filler particles is still able to contribute towards support of higher stress levels. The model presented in Figure 18 should be viewed as suggestive and more work is needed to clarify the exact morphological situation.

Acknowledgments

The help of E. T. Kauchak in sample preparation of P. K. Agarwal in valuable discussions and of J. Alonzo in STEM micrography is highly appreciated. A special acknowledgement is extended to K. B. Wagener whose discussions stimulated this collaborative work.

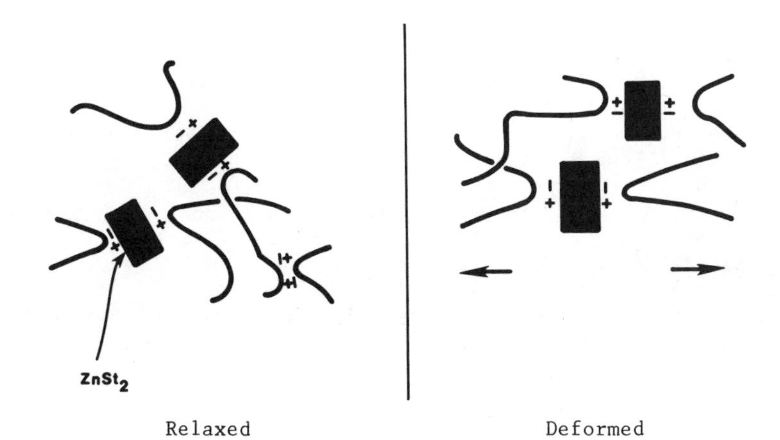

Figure 18. A proposed morphological representation of zinc stearate crystals interacting with Zn-S-EPDM molecules under relaxed and deformed states. Interactions occur between sulfonated sites on the polymeric backbone and polar sites on the surface of the crystals.

Literature Cited

1. Holliday, L., Ed. "Ionic Polymers"; Applied Science Publishers: London, 1975.
2. Eisenberg, A.; King, M. "Ion-Containing Polymers"; Academic Press: New York, 1977.
3. Eisenberg, A., Ed. "Ions In Polymers"; ADVANCES IN CHEMISTRY SERIES No. 187, American Chemical Society: Washington, D.C., 1980.
4. Lundberg, R. D. In "Encyclopedia of Chemical Technology"; Supplement Volume, John Wiley: New York, 1984; pp. 546-573.
5. Lundberg, R. D.; Makowski, H. S. In "Ions In Polymers"; Eisenberg, A., Ed.; ADVANCES IN CHEMISTRY SERIES No. 187, American Chemical Society: Washington, D.C., 1980; Paper No. 2.
6. Makowski, H.; Lundberg, R. D.; Westerman, L.; Bock, J. In "Ions In Polymers"; Eisenberg, A., Ed.; ADVANCES IN CHEMISTRY SERIES No. 187, American Chemical Society: Washington, D.C., 1980; Paper No. 1.
7. Frisch, K. C. In "Polyurethane Technology"; Bruins, P. F., Ed.; Wiley: New York, 1969.
8. Milkovich, R. S. African Patent 280 712, 1963, Holden, G.; Mikkovich, R. U.S. Patents 3 231 635 and 3 265 765, 1966.
9. Coleman, D. J. Polym. Sci. 1954, 14, 15.
10. Makowski, H. S.; Lundberg, R. D. In "Ions In Polymers"; Eisenberg, A. Ed.; ADVANCES IN CHEMISTRY SERIES No. 187, American Chemical Society: Washington, D.C., 1980; Paper No. 3.
11. Canter, N. H.; Buckley, D. J. U.S. Patent 3 847 854, 1974; Bock, J.; Lundberg, R. D.; Phillips, R. R.; Makowski, H. S. U.S. Patent 3 974 240, 1976; and Duvdevani, I.; Bock, J. U.S. Patent 4 151 137, 1979.
12. Duvdevani, I.; Agarwal, P. K.; Lundberg, R. D. Polym. Eng. and Sci. 1982, 22, 500.
13. Canter, N. H. U.S. Patent 3 642 728, 1972.
14. Wood, C. B. Master of Science Thesis, Virginia Polytechnic Institute and State University, Blacksburg, 1982.
15. Agarwal, P. K.; Makowski, H. S.; Lundberg, R. D. Macromolecules 1980, 13, 1679.
16. Wagener, K. B.; Duvdevani, I. "Crystallization and Orientation of Zinc Stearate in Sulfonated EPDM"; presented at the ACS meeting, Houston, March 1980 (unpublished).
17. Schwarzel, F. R.; Bree, H. W.; Nederveen, C. J.; Schwippert, G. A.; Struik, L. C. E.; van der Wal, C. W. Rheol. Acta 1966, 5, 270.
18. Mohajer, Y.; Wilkes, G. L. J. Polym, Sci., Polym. Phys. Ed. 1982, 20(3), 457.
19. Smith, T. L. Trans. Soc. Rheol. 1959, 3, 113.
20. Kerner, E. H. Proc. Phys. Soc. London 1956, 69(B), 808.
21. Guth, E. J. Appl. Phys. 1945, 16, 20.
22. Smallwood, H. J. Appl. Phys. 1944, 15, 758.
23. Agarwal, P. K.; Prestridge, E. B. Polymer 1983, 24, 487.
24. Handlin, D. L. Ph.D. Thesis, The University of Massachusetts, Amherst, 1983.

RECEIVED June 10, 1985

16

Solution Behavior of Metal-Sulfonate Ionomers

R. D. Lundberg[1] and R. R. Phillips[2]

[1]Exxon Chemical Company, Linden, NJ 07036
[2]Corporate Research Science Laboratories, Exxon Research and Engineering Company, Annandale, NJ 08801

Studies on the dilute solution behavior of sulfonated ionomers have shown these polymers to exhibit unusual viscosity behavior in solvents of low polarity. These results have been interpreted as arising from strong ion pair associations in low polarity diluents. Solvents of higher polarity, such as dimethyl sulfoxide and dimethyl formamide induce classic polyelectrolyte behavior in sulfonate ionomers even at very low sulfonate levels. To a first approximation these two behaviors, ion pair interactions or polyelectrolyte behavior, are a consequence of solvent polarity. Intramolecular association of Lightly Sulfonated Polystyrene (S-PS) results in a reduced viscosity for the ionomer less than that of polystyrene precursor at low polymer levels. Inter-association enhances the reduced viscosity of the ionomer at higher polymer concentrations. Isolation of the intra- and inter-associated species of S-PS has been attempted (via freeze drying). A comparison of selected properties reveals significant differences for these two conformations.

During the past 20 years, there have been many publications and patents describing the influence of ionic groups pendant to a hydrocarbon polymer chain on polymer properties (1-4). The presence of as little as 1 mol % of metal carboxylate or metal sulfonate groups can have an especially profound influence on the melt viscosity of polymers whose backbones are comprised largely of ethylene or styrene repeat units. Other physical properties as well as polymer solution behavior are similarly altered (5-7). The influence of these ionic groups on polymer properties is generally attributed to the association of ion pairs resulting in an effective network at sufficiently high ionic levels and high polymer concentration.

0097-6156/86/0302-0201$06.00/0

The nature of the ion pair aggregates has been examined in some detail in publications by several authors. Specific models have been proposed to describe the nature of the ion pair aggregates (8, 9), and a number of studies have been conducted to elucidate the size and morphology of these aggregates experimentally (1, 10). While there has been some ambiguity concerning the nature of these ion-rich phases, there is substantial evidence to suggest that they do exist. Their size and molecular arrangement are still open to question.

Recent studies in our laboratories have been concerned with the physical properties of sulfonated ionomers such as sulfonate ethylene/propylene/ethylidene norbornene terpolymers (4), or lightly sulfonated polystyrene (S-PS)(11). These ionomers exhibit pronounced ion pair association (at sulfonate levels $\geq$ 15 milli-equivalents/100 g polymer) to a degree that they appear crosslinked covalently. These interactions can be dissipated by the addition of a polar additive, thereby showing that such associations are indeed physical and do not arise due to covalent crosslinking.

Several studies (6, 13) of the solution behavior of sulfonate ionomers have provided additional insight on the nature of the ion pair aggregation. The polarity of the solvent environment has been shown to have a major influence on the dilute solution behavior of these polymers. In the course of these studies it has been observed with selected systems that both melt viscosity values and solution behavior can vary according to the history of sulfonate ionomers. This study provides some data and provides one rationale for such differences.

EXPERIMENTAL

The synthesis of lightly sulfonated polystyrene or S-PS has been described in several publications and patents (5, 6). In the current study only the sodium salt of S-PS has been investigated. The details of specific isolation procedures have been described in those references and are summarized in the Results section. The melt viscosity measurements were conducted in a melt index rheometer under conditions which have been previously described. The starting polystyrene employed in these studies was a commercial homopolymer designated as Styron 666 marketed by the Dow Chemical Company. This polymer had an intrinsic viscosity of 0.80 in toluene at 25°C, a number average molecular weight of 106,000 (as established by gel permeation chromatography). Sulfonation/neutralization reactions were conducted as described in various publications (4). On the basis of available data from GPC and intrinsic viscosity measurements, we conclude that the sulfonation/neutralization process has no significant effect on the backbone molecular weights at least for polystyrenes having sulfonation levels below about 3 mole percent.

In one system a narrow molecular weight distribution product (obtained from Polysciences) was used. This product had M_n = 220,000 and M_w/M_n = 1.06. The results described were those obtained with commercial polystyrene unless otherwise indicated.

The dilute solution viscosity measurements were conducted using Ubbelohde viscometers generally conducted at 25 $\pm$ 0.05°C in a thermostated bath. The reduced viscosity or viscosity number [defined as $(\eta-\eta_o)/\eta_o c$, where η is the viscosity of the polymer solution, η_o is the viscosity of the solvent (or mixed solvent), and c is the concentration of polymer in g/100 mL] was calculated for each solution measured.

RESULTS

I. Dilute Solution Behavior of S-PS

Lightly sulfonated polystyrene is soluble in mixed solvent systems, such as xylene containing low levels of alcohols, or in moderately polar solvents. In low polarity solvents the viscosity of such ionomer solutions can be substantially higher than polystyrene of comparable molecular weight due to ion pair association at concentrations >1% as shown in Table I.

Table I. Viscosity (centipoise) of 1.7 Mol % S-PS and Polystyrene in Solvent Mixture (Xylene + 3% Hexanol)

1.7 Mol % SPS (Na salt)		Polystyrene	
Concentration, %	Viscosity, cp	Concentration, %	Viscosity, cp
		23	275
		17.5	102
		10	24
5	2,066	5	6.2
3	162	3	3.3
2	6	2	2.0
1	1.3	1	1.2
0.5	0.87	0.5	0.9
0.25	0.72	0.25	0.76
0.125	0.68	0.125	0.70

The effect of polar additives is to moderate the ion pair associations. This influence of polar cosolvents can be described as a specific solvation of the metal cation groups by more polar species and can be viewed as to a first approximation as a simple equilibrium (6).

$$\mathrm{ROH} + (\mathrm{P} - \mathrm{SO_3^-Na^+})_n \rightleftharpoons {}_n(\mathrm{ROH:}\ \mathrm{Na^+SO_3^-} - \mathrm{P})$$

While this equilibrium ignores the polymer backbone and the majority of the solvent system, it does predict many characteristics of metal sulfonate ionomers in hydrocarbon solution.

Moderately polar solvents such as tetrahydrofuran are also effective for SPS up to a sulfonate level of about 5 mole % in the case of the sodium salt (7). A comparison of the reduced viscosity-concentration profiles of S-PS at various sulfonate levels is compared with that of the PS precursor (Figure 1). It is evident that there is a regular progression in Reduced Viscosity as a function of sulfonate content with increasing values at higher sulfonate levels observed at higher concentrations (~2%). Conversely the higher sulfonated levels in SPS lead to lower viscosity values at low polymer levels. This behavior is readily rationalized as being a consequence of intra-associations at lower polymer levels and inter-associations at the higher polymer concentrations. Analogous behavior is observed in hydrocarbon solvents or those containing low levels of polar cosolvents.

One implication of Figure 1 relates to polymers containing higher levels of metal sulfonate groups. It might be expected that such polymers might provide homogeneous gels at high polymer levels and yet display phase separation upon dilution. Such behavior is often observed for metal sulfonate ionomers in hydrocarbon or hydrocarbon-alcohol solvent systems.

Few studies have been conducted heretofore on sulfonated ionomers in solvents which can be considered relatively polar, as defined by a high dielectric constant. A recent study (13) on acrylonitrile-methallyl sulfonate copolymers in dimethyl-formamide is a notable exception. S-PS is readily soluble in a wide variety of solvents, some of them exhibiting rather high values of dielectric constant, such as dimethylformamide (DMF) or dimethylsulfoxide (DMSO). The reduced viscosity-concentration behavior of sulfonated polystyrene is markedly different in polar solvents from that in nonpolar-solvent systems. Typically there is a marked upsweep in reduced viscosity at low polymer concentrations and clearly a manifestation of classic polyelectrolyte behavior. (7)

Typical polyelectrolyte behavior has been observed in several solvents of moderate to high polarity as described in Table II. It is clear that in low polarity solvents ion-pair association prevails, while with high polarity solvents polyelectrolyte behavior is observed. With solvents of intermediate polarity either behavior can be observed suggesting specific solvation effects.

Table II. Relation of Dielectric Constant and Dilute-Solution Behavior (7)

Solvent	Dielectric Constant	Polyelectrolyte Behavior?
Dimethylsulfoxide	46.7	Yes
Dimethylformamide	36.7	Yes
Cyclohexanone	18.2	No
Ethylene glycol monomethyl ether (2-methoxyethanol)	16.9	Yes
Tetrahydrofuran	7.6	No
Dioxane	2.2	No

Source: Reproduced with permission from Ref. 7. Copyright 1982 J. Polym. Sci., Polym. Phys. Ed.

While solvents such as THF and Dioxane exhibit ion pair association with sulfonate ionomers, the addition of more polar cosolvents can have a marked effect on the reduced viscosity-concentration profiles. For example, Figure 2 illustrates the influence of varying

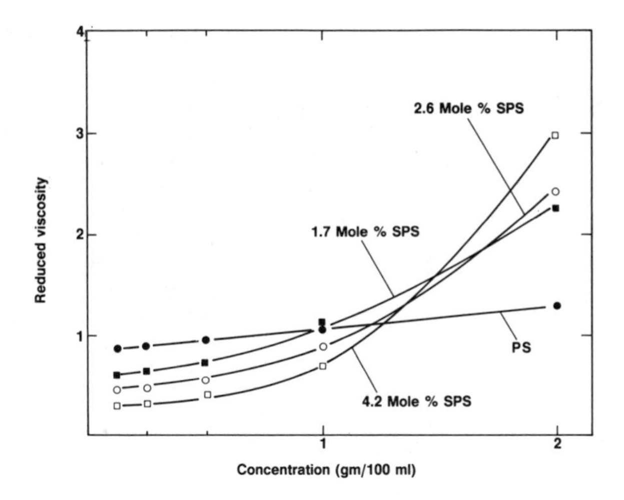

Figure 1. Reduced viscosity-concentration relationships of PS and S-PS of different sodium styrene sulfonate levels in tetrahydrofuran. Reproduced with permission from Ref. 7. Copyright 1982 J. Polym. Sci., Polym. Phys. Ed.

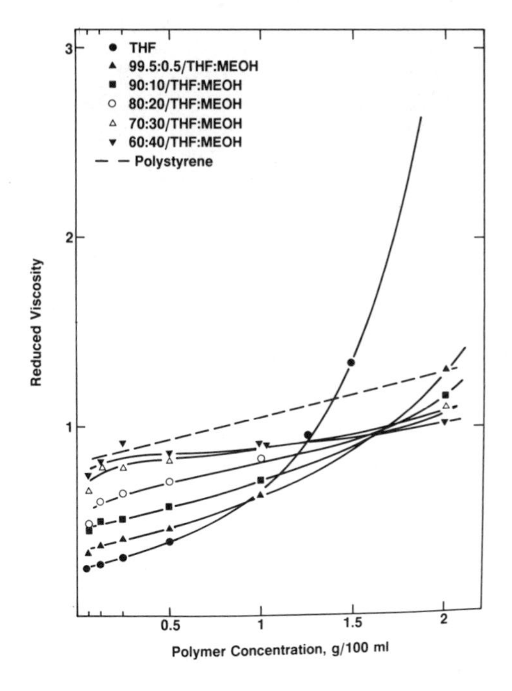

Figure 2. Reduced viscosity-concentration relationships for S-PS in tetrahydrofuran containing various levels of methanol. The S-PS contained 4.2 mole percent sodium sulfonate groups.

amounts of methanol on S-PS of 4.2 mole percent of sodium sulfonate content. At methanol contents of 40 volume percent the intrinsic viscosity of S-PS is quite similar to that of the unfunctionalized polystyrene. To a first approximation, Figure 2 at varying cosolvent levels is similar to Figure 1 with varying sulfonate levels. In effect both curves display the influence of varying degrees of ionic association for ionic groups pendant to a common backbone.

The influence of small amounts of water as a cosolvent for THF solutions is shown in Figure 3. At water contents of 3% or less the reduced viscosity-concentration profiles are similar to Figure 2. However, increasing water levels induces an upsweep in reduced viscosity at low polymer concentrations typical of polyelectrolyte behavior. This behavior has been observed in other mixed solvent systems and is clearly a consequence of specific cation solvation effects. Sodium^{23}-NMR studies have shown that water solvates sodium cations at least ten times more strongly at similar cosolvent levels and the data in Figures 2 and 3 are consistent with those findings.

II. Relation of Solution Properties and Bulk Polymer Viscosity

To a very rough approximation the concentration where reduced viscosity-concentration plots of S-PS and PS intersect in solvents where ion pair associations prevail can be viewed as that concentration where polymer coils overlap, corresponding to C* (Figure 4). In effect, these studies suggest that S-PS exists in two different conformations at different polymer concentrations. Furthermore, the very strong associations which can prevail in these systems could preclude a rapid interchange between the two species under certain conditions. It can be argued that _if_ polymer conformation will control polymer properties for ionomers, then a difference in physical property behavior might be expected for a material isolated at low polymer concentrations as compared to the same material isolated from high polymer concentrations (i.e., >> C*).

Typically conventional polymer isolation procedures do not permit product isolation within a specific concentration region; however, under certain conditions the operation of freeze drying ionomer solutions would be expected to offer this option. Thus, freeze drying of S-PS at two different concentrations (0.3 and 4.0 weight percent polymer) from appropriate solvents should offer two different polymer species. This hypothesis, of course, assumes that once solutions of this polymer are frozen, the polymer conformations will be "locked in" at those concentrations, or the normal changes which might occur under other conditions of polymer isolation will be minimized.

A practical freeze-drying approach with S-PS requires a suitable solvent which freezes at a relatively high temperature, yet will sublime at a rapid rate. P-Dioxane meets these criteria and also offers the required ion pair association behavior as shown in Figure 4.

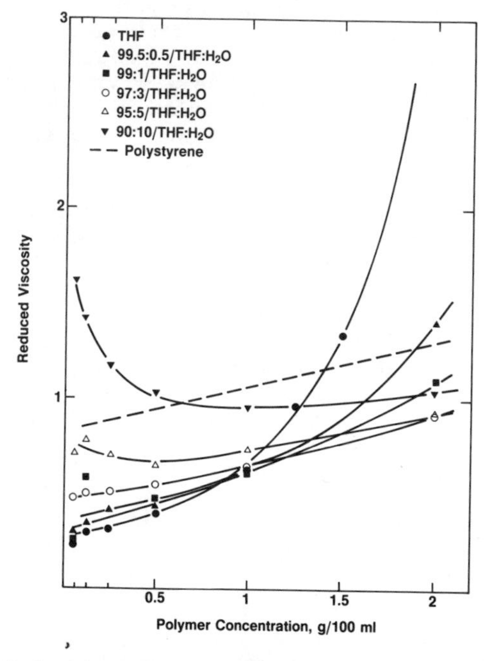

Figure 3. Reduced viscosity-concentration relationships for S-PS in tetrahydrofuran containing various levels of water. The S-PS contained 4.2 mole percent sodium sulfonate groups.

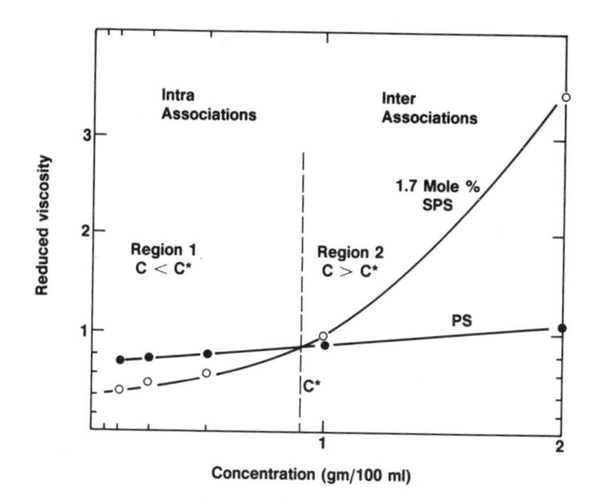

Figure 4. Reduced viscosity-concentration relationships for S-PS (1.7 mole % sodium salt) in Dioxane. Reproduced with permission from Ref. 13. Copyright 1984 J. Polym. Sci. Polym. Lett. Ed.

In a series of experiments, S-PS (1.7 mole % sodium sulfonate) was freeze dried at two polymer concentrations from dioxane solvent (0.3% and 4.0%). Characterization was then conducted on the isolated products and on the initial S-PS. Characterization of the samples was attempted by comparing the melt viscosities of the isolated products and by solubility behavior. Similarly, PS was freeze-dried under the same conditions and characterized in the same way.

The melt viscosities of the materials were measured employing a melt rheometer as previously described (11) where the melt viscosities were obtained at constant shear stress as a function of time. Those results are summarized in Table III. The polymer isolated by freeze drying from the more dilute solution exhibits an initial melt viscosity which is less than the material derived from the concentrated solution (or the original stock S-PS) by a factor of 10. Directionally it appears that the viscosity of the product isolated from the dilute solution increases significantly with time.

The product isolated at dilute concentration exhibits other properties different from that of the material isolated at higher concentration. For example, conventional S-PS (1.7%) will form a homogeneous gel in xylene at concentration >3%, but will phase separate to form a gel phase in more dilute solutions, especially <1%. This behavior has been observed with a number of sulfonate ionomers.

S-PS, isolated by freeze drying from dilute solution, invariably is more soluble under such conditions than S-PS isolated via freeze drying at 4% concentration. This behavior is summarized in Table III below.

Table III. Melt Viscosity and Solubility Behavior of S-PS (Na Salt) in Xylene after Freeze Drying from Dioxane

(Sulfonate Level)	Concentration of Freeze-Dried Solution	Melt Viscosity 220°C, Poise x 10^{-5}	% Soluble in Xylene 1g/100 cc
A (1.7 Mole %)	4	700	37
	.3	70	89
*B (1.98 mole %)	4	150	63.7
	.3	2.6	90.2
C (2.02 mole %)	4	61.5	57.3
	.3	2.4	81.9

* Starting polystyrene was a narrow molecular weight distribution material from Polysciences.

Source: Reproduced with permission from Ref. 13. Copyright 1984 J. Polym Sci., Polym Lett. Ed.

These data indicate that the intramolecular interactions which prevailed in dilute solution and which remained "locked in" the solid polymer upon freeze drying are still maintained upon dissolution in xylene. The resulting "solutions" in xylene remain unchanged at ambient conditions over periods of months. It appears that equilibration of inter/intra associations readily occur upon dissolution or dilution in dioxane with S-PS at 1.7% sulfonation levels. However, with the same sample dissolved in a solvent of lower polarity, such as xylene, equilibration upon dilution either does not occur or is very slow. The addition of cosolvents such as alcohols expedites this equilibration. At different sulfonate levels, with different cations or different backbone molecular weights the situation may be different.

We have observed that polystyrene itself shows no significant change in melt viscosity when isolated under the conditions described above, as would be expected. (13)

These observations suggest that sample preparation of sulfonate ionomers could be very important in determining physical properties of such systems. There are several qualifying comments that should be made.

The data relating to influence of polymer isolation on polymer melt viscosity or solution behavior has been developed solely in diluents where ion pair association prevails.

Polar solvents such as dimethylformamide, dimethylsulfoxide, and tetrahydrofuran-water mixtures behave differently in that polyelectrolyte behavior is observed at extreme dilution for sulfonate ionomers; therefore, the behavior described above does not apply directly to these solvent systems.

Thus far we have not succeeded in the isolation of an ionomer free of impurities from a solvent favoring polyelectrolyte behavior where its solution behavior can be compared to that in Table II. Currently such studies are in progress.

Literature Cited

(1) W. J. MacKnight and T. R. Earnest, J. Polym. Sci. Macromol. Rev., 16, 41 (1981).

(2) A. Eisenberg and M. King, "Ion Containing Polymers", Academic, New York 1977.

(3) A. Eisenberg and M. Navratil, Macromolecules, 7, 90, (1974).

(4) H. S. Makowski, R. D. Lundberg, L. Westerman, and J. Bock "Ions in Polymers", ACS Monogr. 187, A. Eisenberg, Ed., Am. Chem. Soc., Washington, D.C., 1980, Chap. 1.

(5) R. D. Lundberg, Polym. Prepr. Am. Chem. Soc. Div. Polym. Chem., 19(1), 455 (1978).

(6) R. D. Lundberg and H. S. Makowski, J. Polym. Sci. Polym. Phys. Ed., 18, 1821 (1980).

(7) R. D. Lundberg and R. R. Phillips, J. Polym. Sci. Polym. Phys. Ed., 20, 1143 (1982).

(8) R. Longworth and D. J. Vaughan, Polym. Prepr. Am. Chem. Soc. Div. Polym. Chem., 9, 525 (1968).
(9) W. J. MacKnight, W. P. Taggart, and R. S. Stein, J. Polym. Sci. Polym. Symp., 45, 113 (1974).
(10) M. Pineri, C. Meyer, A. M. Levelut, and M. Lambert, J. Polym. Sci. Polym. Phys. Ed., 12, 115 (1974).
(11) R. D. Lundberg, H. S. Makowski and L. Westerman, ACS Monograph 187, 1980, p. 67.
(12) Unpublished Results; R. D. Lundberg, M. T. Melchior, B. Hudson of Exxon Research & Eng. Co.
(13) R. D. Lundberg and R. R. Phillips, J. Polym. Sci., Polym. Lett. Ed., 22, 377 (1984).

RECEIVED January 24, 1986

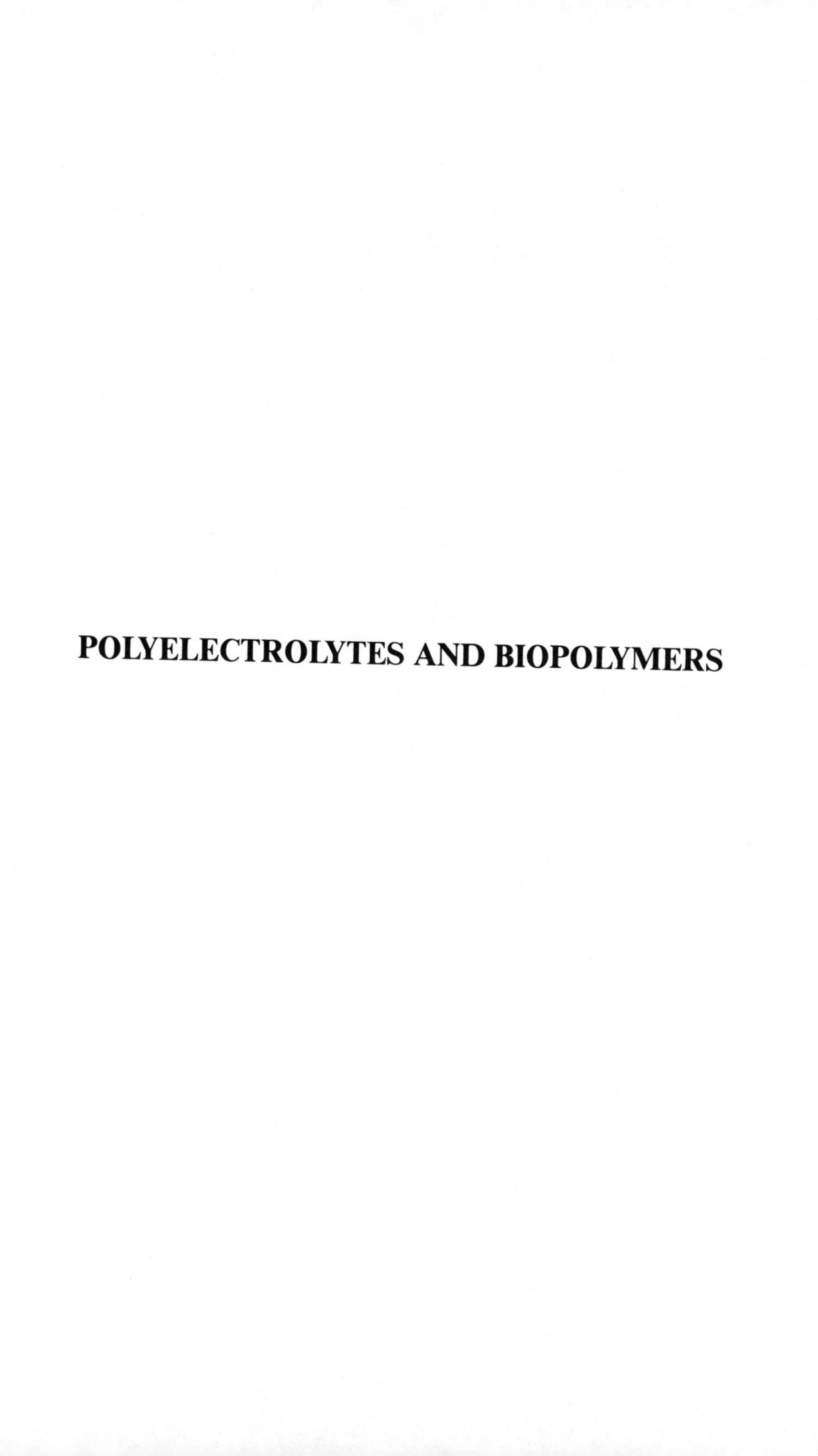

POLYELECTROLYTES AND BIOPOLYMERS

17

Use of the Poisson–Boltzmann Equation To Predict Ion Condensation Around Polyelectrolytes

Bruno H. Zimm

Department of Chemistry, B–017, University of California, San Diego, La Jolla, CA 92093

We summarize recent work showing that condensation can be derived as a natural consequence of the Poisson-Boltzmann equation applied to an infinitely long cylindrical polyelectrolyte in the following sense: Nearly all of the condensed population of counter-ions is trapped within a finite distance of the polyelectrolyte even when the system is infinitely diluted. Such behavior is familiar in the case of charged plane surfaces where the trapped ions form the Gouy double layer. The difference between the plane and the cylinder is that with the former all of the charge of the double layer is trapped, while with the latter only the condensed population is trapped.

Manning's condensation theory (1) describes the behavior of small ions around a long, highly charged polyelectrolyte by postulating that there are two populations of counter-ions, one normal, and one "condensed". If we have counter-ions with valence z, the fraction, F_M, of these ions in the condensed population is

$$F_M = 1 - 1/z\xi$$

ξ, called the linear charge-density parameter of the polyelectrolyte, is equal to $e^2/4\pi\varepsilon_0 DkTb$, where e is the electron charge, ε_0 the capacitivity of the vacuum, D the dielectric constant of the solvent, kT as usual, and b is the axial spacing of the (univalent) charges on the polyelectrolyte. The above formula holds if the charge density on the polyelectrolyte is high enough so that ξ is greater than $1/z$; otherwise there is no condensed population. This theory has scored a number of remarkable successes, particularly in regard to DNA as the polyelectrolyte. (For reviews, see Manning (2) and Anderson and Record (3).)

For a long time, however, the literature did not give a clear picture of how these condensed ions were distributed in space; there was some disagreement between Manning's own picture, in which the

0097–6156/86/0302–0212$06.00/0

condensed ions occupied a fixed and rather small volume around the polyelectrolyte (4), and solutions of the Poisson-Boltzmann (Gouy-Chapman) equation (5,6), in which the condensed ion cloud was shown to be diffuse and to extend out to distances that approached infinity at infinite dilution. Recently Marc Le Bret and I (7,8) have been able to remove most of the sources of disagreement by examining in detail the solutions of the Poisson-Boltzmann equation with the polyelectrolyte modeled as an infinitely long charged cylinder. In the solutions of this equation at infinite dilution we find a "bound", or "condensed", cloud of counter-ions, a cloud defined by the fact that essentially all of it remains within a finite distance of the poly-ion even at infinite dilution. The number of ions in the cloud is exactly that given by Manning's condensation formula. (We put the words "bound" and "condensed" in quotation marks because they have been used in many different senses in the past; here they are intended only in the sense just defined. We are not prepared to argue whether they are the best words for the purpose.) The details of this calculation have already been published (7,8), so we present here only a brief description and the main conclusions.

The system that we consider is an infinitely long cylindrical poly-ion enclosed in an outer concentric cylindrical container filled with solvent and counter-ions of valence z but with no added salt; this is one case in which analytic, as opposed to numerical, solutions of the Poisson-Boltzmann equation are available. Numerical solutions for the case where added salt is present show much the same picture, however, so this limiting case with counter-ions only is still of general interest. The Poisson-Boltzmann equation for this system was solved long ago (9,10).

With this solution, which gives the concentration of counter-ions at any position, we can proceed in the following way to find the condensed fraction of counter-ions. We consider the case where the linear charge density on the poly-ion is high, so that ξ is greater than $1/z$. From the solution we can calculate the radius, $r(F)$, measured from the cylinder axis, that contains a given fraction, F, of the counter-ions, and we can find the limiting value of this radius as the average concentration of counter-ions goes to zero, which happens when the outer cylinder is made infinitely large. When we do this we find that the behavior of $r(F)$ is very different depending on whether F is greater or less than the F_M of Eq. (1). When $F > F_M$ then

$$r(F) \propto c^{-1/2}$$

which goes to infinity as c, the average concentration, goes to zero, but when $F < F_M$ then

$$r(F) = a \exp[F/((z\xi-1)(F_M-F))]$$

which remains finite. Thus if we take any fraction less than F_M of counter-ions, this fraction remains within a finite radius of the poly-ion at infinite dilution; these are the so-called condensed counter-ions. These counter-ions form a cloud around the poly-ion, all of the cloud lying within a finite radius of the poly-ion. In contrast, ions in excess of the fraction F_M dilute away to infinity at infinite dilution.

Away from infinite dilution, and even when some added salt is present, the situation of the condensed cloud is not much different. At finite average concentrations the cloud moves in somewhat closer to the poly-ion, and of course the non-condensed ions are present also. The concentration of counter-ions at the surface of the poly-ion is remarkably high. For example, with a poly-ion with the approximate characteristics of DNA ($\xi = 4$ and a radius of 1.25nm), the concentration of univalent counter-ions at the surface stays near 3 molar even at infinite dilution, and this concentration is remarkably insensitive to the overall average concentration unless the latter exceeds 1 molar.

The Plane also Shows "Condensation"

Condensation in the above sense, a cloud of ions remaining within a finite distance even at infinite dilution, is not unique to the infinitely long charged cylinder, although the phenomenon is not usually known by that name. The mobile ions of the double layer next to a charged infinite plane surface behave in the same way. The solution of the Poisson-Boltzmann equation for this case has been known even longer than for the cylinder (11); from it we can calculate the behavior of the mobile ions as their average concentration approaches zero. Let Q be the amount of fixed charge per unit area on the solid planar surface, and let l_B be the so-called Bjerrum length defined by

$$l_B = e^2/(4\pi\varepsilon_0 DkT)$$

and let X be a distance measured from the solid surface to a plane in the liquid. This plane and the solid surface enclose between them a certain fraction, F, of the total mobile-ion charge, which total charge per unit area of surface is equal to and of opposite sign to Q. As with the cylinder, we can calculate X(F), the distance corresponding to a given F, and take the limit at infinite dilution. This limit is (8) (for a symmetrical electrolyte)

$$X(F) = \frac{2e}{z l_B Q} \frac{F}{1-F}$$

which has a finite value for any F less than unity. Thus almost all of the mobile-ion charge stays within a finite distance X of the solid plane surface at infinite dilution, and this charge satisfies the same definition of "condensed" as the fraction F_M of the cylinder case. The only difference is that in the case of the plane effectively all of the charge is "condensed", while in the case of the cylinder only the fraction F_M, which has a value anywhere between 0 and 1 depending on the linear charge density on the cylinder, is "condensed".

In conclusion, we can say that the Poisson-Boltzmann description of the condensed population of mobile ions near a charged cylinder is similar to the Poisson-Boltzmann description of the mobile ions in the Gouy diffuse double layer at a charged planar surface, a description that has been well known for a long time. In both cases the ions are "bound", or "condensed", in the sense that they cannot dilute away as the volume of the system is expanded indefinitely.

Literature Cited

1. Manning, G.S. J. Chem. Phys. 1969, 51, 924-933, 3249-3252.
2. Manning, G.S. Q. Rev. Biophys. 1978, 11, 179-246.
3. Anderson, C.F.; Record, M.T. Ann. Rev. Phys. Chem. 1982, 33, 191-222.
4. Manning, G.S. Biophys. Chem. 1977, 7, 95-102.
5. MacGillivray, A.D. J. Chem. Phys. 1972, 56, 81-83, 83-85.
6. Gueron, M.; Weisbuch, M. Biopolymers 1980, 19, 353-382.
7. Le Bret, M.; Zimm, B.H. Biopolymers 1984, 23, 287-312.
8. Zimm, B.H.; Le Bret, M. J. Biomolec. Structure and Dynamics 1983, 1, 461-471.
9. Alfrey, T.; Berg, P.W.; Morawetz, H. J. Polymer Sci. 1951, 7, 543-547.
10. Fuoss, R.M.; Katchalsky, A.; Lifson, S. Proc. Nat. Acad. Sci., U.S. 1951, 37, 579-589.
11. Verwey, E.J.W.; Nissen, K.F. Phil. Mag. 1939, (7)281, 435-446.

RECEIVED June 5, 1985

18

Dynamics of Macromolecular Interactions

Stuart A. Allison[1,3], J. Andrew McCammon[1], and Scott H. Northrup[2]

[1]**Department of Chemistry, University of Houston, Houston, TX 77004**
[2]**Department of Chemistry, Tennessee Technological University, Cookeville, TN 38505**

The rates of important processes in macromolecular solutions are often influenced or controlled by the binary diffusional encounter frequency of reactants. Two examples are the growth of polymer chains and the binding of ligands to receptors. Calculation of reaction rates in such systems generally requires consideration of such factors as anisotropic Coulombic and hydrodynamic interactions between reactants, and orientation dependent reactivity of the collision partners. A computer simulation approach has been derived that allows detailed bimolecular reaction rate constant calculations in the presence of these and other complicating factors. In this approach, diffusional trajectories of reactants are computed by a Brownian dynamics procedure; the rate constant is then obtained by a formal branching anaylsis that corrects for the truncation of certain long trajectories. The calculations also provide mechanistic information, e.g., on the steering of reactants into favorable configurations by electrostatic fields. The application of this approach to simple models of enzyme-substrate systems is described.

The frequency with which two reactive species encounter one another in solution represents an upper bound on the bimolecular reaction rate. When this encounter frequency is rate limiting, the reaction is said to be diffusion controlled. Diffusion controlled reactions play an important role in a number of areas of chemistry, including nucleation, polymer and colloid growth, ionic and free radical reactions, DNA recognition and binding, and enzyme catalysis.

Smoluchowski and Debye investigated the problem of diffusion controlled reactions between uniformly reactive spheres in the absence (1) and presence (2) of centrosymmetric Coulombic forces. Since these pioneering works, there has been a proliferation of theoretical studies based on more refined models. These have considered the inclusion of hydrodynamic interaction, (3-4) solvent

[3]Current address: Department of Chemistry, Georgia State University, Atlanta, GA 30303

0097-6156/86/0302-0216$06.00/0

caging effects, (5) concentration effects, (6) orientation dependence of reactivity on one or both species, (7-10) internal-configuration-dependent reactivity, (11-13) and noncentrosymmetric direct forces (14). A number of excellent reviews discuss these and other factors in more detail (6, 15-16). Perhaps the most advanced analytical-numerical methods are those based on the formalism of Wilemski and Fixman (17-18) and extended by other investigators, (19-21) and also the numerical methods of Zientra, Freed, and coworkers (22). These methods have been particularly useful in intramolecular reaction processes such as ring closure in chain molecules (20) and protein domain coalescence (22).

Here, a new method is described in which biomolecular rate constants are determined by a relatively simple simulation procedure (23). This method is sufficiently general to model systems of arbitrary configurational complexity, arbitrary inter- and intramolecular forces, and allows for inclusion of hydrodynamic interaction. When a variety of interactions are present between the reactive species, there is probably little hope of obtaining analytical rate constants at a detailed level and recourse to simulation methods becomes necessary. In this work, the role of local and long range electrostatic forces on diffusion controlled reactions is of primary interest. Anisotropic reactivity and inclusion of hydrodynamic interaction are factors that are studied as well. In the next section, we explain how a rate constant can be derived from the simulation of a large number of trajectories and how a trajectory is computed. In the section on applications, the methodology is applied to three progressively more complex model systems. In the first model (two reactive spheres), the effects of centrosymmetric coulombic forces, anisotropic reactivity, and hydrodynamic interactions are considered. In the second model (a dimer reacting with a sphere), it is shown that non-centrosymmetric Coulombic interactions can act to "steer" the dimer into a favorable orientation for reaction with the sphere. Similar behavior is also observed in the third model, designed to represent the reaction between the enzyme superoxide dismutase and the substrate superoxide. In the final section, we summarize the results of the preceeding section and briefly discuss future applications.

Methodology

For diffusion-influenced bimolecular reactions, one is ordinarily most interested in obtaining a bimolecular rate constant k in order to make contact with experimental studies. To obtain a rate constant by a simulation procedure, one would, in principle, need to simulate a large ensemble of reactant pairs diffusing from large separation to the reaction surface. However, the need to simulate reactant displacements in an infinite domain was obviated by a recent derivation connecting k to a recombination probability β for a pair of reactants diffusing in a finite domain. As depicted in Figure 1, a hypothetical sphere of radius b divides the relative separation space $\underset{\sim}{r}$, into an outer region ($r > b$) and an inner region ($r < b$). The radius b is chosen sufficiently large so that i) interparticle direct and hydrodynamic forces are centrosymmetric to a good approximation at $r=b$, and ii) the ensemble reactive flux through the $r = b$ surface is

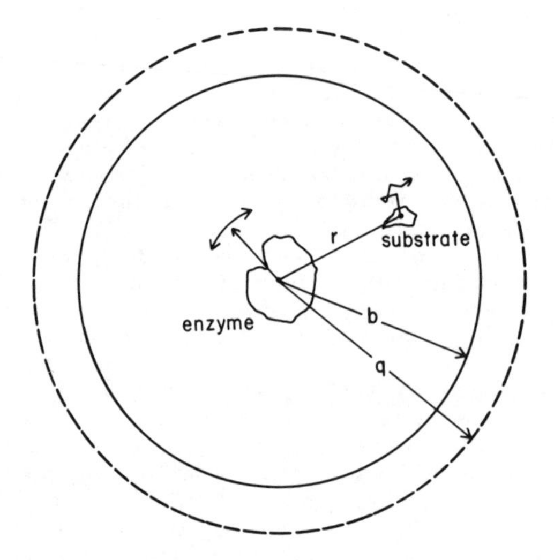

Fig. 1. Schematic Illustration of the Method. Trajectories are started at b, which defines the separation of an anisotropic inner region ($r<b$) and an isotropic outer region ($r>b$). Trajectories are terminated upon reaction or when $r>q$ (23).

isotropic. This second condition can be relaxed to yield improved computational efficiency (42). Under steady state conditions, k is given by

$$k = k_D(b)p \tag{1}$$

where p is the probability that the reactant pair, starting at initial separation r = b, will react rather than diffuse apart and $k_D(b)$ is the familiar Debye rate constant for pairs with r initially >b to first achieve a separation r = b. Because of the restrictions placed on b, $k_D(b)$ can be determined analytically and is given by (5)

$$k_D(b) = \left(\int_b^\infty dr\left[\frac{\exp\,[u(r)/k_BT]}{4\pi r^2 D(r)}\right]\right)^{-1} \tag{2}$$

where u(r) is the (centrosymmetric) potential of mean force and D(r) is the relative diffusion constant discussed in more detail later.

To avoid the problem of reactants diffusing to large distances in the determination of p, trajectories are terminated if r exceeds some cutoff distance q depicted in Figure 1. What is actually determined in a simulation over many trajectories is a recombination probability, β. Since it is possible that a trajectory which reaches separation r ⩾ q would react if not terminated, p and β are not equal. Using branching arguments, however, it is possible to correct β to account for this discrepancy (23). In the special case where all reactive surface collisions lead to a reaction

$$k = \frac{k_D(b)\beta}{1-(1-\beta)\Omega} \tag{3}$$

$$\Omega = k_D(b)/k_D(q) \tag{4}$$

The more general result in which only a fraction of reactive surface collisions lead to reaction is given elsewhere (23).

In order to simulate the dynamical trajectories of a model system, the Langevin equations of motion are integrated taking discrete time steps (24-29). Since it is the comparatively slow, long-range relative motions of reacting species that are of primary interest here, highly damped Langevin or Brownian dynamics is the most relevant. A number of Brownian dynamics algorithms are available, (24-29) but in this work, the algorithm of Ermak and McCammon is used (24). The interacting particles are modelled as spheres or arrays of spherical subunits. If the initial position of subunit i is $\tilde{r}_i^0$ in a space fixed reference frame, its position after a time step of duration Δt is

$$\mathbf{r}_i = \mathbf{r}_i^o + \Delta t (k_B T)^{-1} \sum_{j=1}^{N} \mathbf{D}_{ij}^o \cdot \mathbf{F}_j^o + \mathbf{S}_i(\Delta t) \tag{5}$$

where k_B is Boltzmann's constant, T is the absolute temperature, $\mathbf{F}_j^o$ is the initial force acting on subunit j excluding stochastic (solvent) forces and, if present, forces of constraint. $\mathbf{S}_i$ is a vector of Gaussian random numbers of zero mean and variance-covariance

$$\langle \mathbf{S}_i \mathbf{S}_j \rangle = 2\, \mathbf{D}_{ij}^o \Delta t \tag{6}$$

The components of $\mathbf{S}_i$ represent stochastic displacements and are obtained using the multivariate Gaussian random number generator GGNSM from the IMSL subroutine library (30). $\mathbf{D}_{ij}^o$ is the initial hydrodynamic interaction tensor between subunits i and j. Although the exact form of $\mathbf{D}_{ij}^o$ is generally unknown, it is approximated here using the Oseen tensor with slip boundary conditions. This representation has been shown to provide a reasonable and simple point force description of the relative diffusion of finite spheres at small separations (31). In this case, one has

$$\mathbf{D}_{ij} = k_B T \left[\frac{\delta_{ij}}{4\pi\eta a_i} \mathbf{I} + (1-\delta_{ij}) \mathbf{T}_{ij} \right] \tag{7}$$

where δ_{ij} is the Kronecker delta, $\mathbf{I}$ is the identity matrix, $\mathbf{T}_{ij}$ is the Oseen tensor

$$\mathbf{T}_{ij} = \frac{1}{8\pi\eta R} \left(\mathbf{I} + \frac{\mathbf{r}_{ij}\mathbf{r}_{ij}}{r_{ij}^2} \right) \tag{8}$$

$$R = \begin{cases} a_i + a_j & r_{ij} < a_i + a_j \\ r_{ij} & r_{ij} \geqslant a_i + a_j \end{cases} \tag{9}$$

and a_i is the radius of subunit i. In some simulations, hydrodynamic interaction (HI) between the reacting species is ignored. In those cases, the approximation is made that $\mathbf{D}_{ij} = D_i \delta_{ij} \mathbf{I}$ where $D_i = k_B T/4\pi\eta a_i$ or in other words, $\mathbf{T}_{ij}$ is set equal to zero in Eq. 7. Since the neglected term falls off as r_{ij}^{-1}, this approximation is expected to work reasonably well when i and j are very far apart. Subsequently, when we speak of the case of "no hydrodynamic interaction" (NHI) we shall be referring to this approximation. In the model of the monomer target interacting with a dimeric ligand, discussed in section on sphere and dumbell dimer, intramolecular HI between the subunits of the dimer is retained even though intermolecular HI is ignored in particular NHI simulations.

If forces of constraint are present, as in the case of the monomer-dimer study where the distance between the dimer subunits is fixed, displacement correction vectors must be added to Eq. (5) in order to enforce the constraints. Enforcing constraints is a troublesome problem in both molecular (32) and Brownian (26, 33) dynamics. Nonetheless, they can be enforced in a rigorous manner. Where needed in this work the SHAKE - HI algorithm described and implemented elsewhere, (33) is used to enforce constraints. When hydrodynamic interaction is present, it turns out that displacement correction vectors must be added to unconstrained as well as constrained subunits. In the monomer dimer studies with HI, for example, a displacement correction vector must be applied to the monomer when the constraint between dimer subunits is enforced.

Applications

Two Spheres. The steady state diffusion controlled rate constant for two uniformly reactive spheres interacting via a centrosymmetric potential of mean force can be solved numerically and in special cases analytically as given by Eq. 2. For a potential of mean force of zero and no hydrodynamic interaction (NHI), Eq. 2 reduces to the Smoluchowski result (1)

$$k_D(b) = 4\pi D_{rel} b \tag{10}$$

where b is the center-to-center distance at which the spheres spontaneously react and $D_{rel} = D_1 + D_2$ where D_1 and D_2 are the translational diffusion constants of the individual spheres. When HI is included, D(r) is given by (6)

$$D(r) = D_{rel}\hat{r}\cdot[\underset{\approx}{I} - \frac{2k_BT}{D_{rel}}\underset{\approx}{T}]\cdot\hat{r} \tag{11}$$

where $\hat{r}$ is the unit relative displacement vector between the spheres and $\underset{\approx}{T}$ is approximated using Eq. 8.

To test the simulation method, we first studied uniformly reactive spheres under conditions such that the simulations can be compared directly to analytic results. Calculations were then carried out for interacting spheres with anisotropic reactivity. One of the spheres was assumed to be reactive only over half its surface whereas the remaining sphere was uniformly reactive. This shall be called the hemisphere model. The results of the simulations involving two reactive spheres are summarized in Table I. Sphere radii of $a_1 = a_2 = 0.5$ Å with a "reaction radius" of 1 Å were used throughout. In addition to studying the effect of hydrodynamic interaction, effects of direct forces were also considered using simple Coulomb and screened Coulomb interactions. In the case of the hemisphere model, several simulations were carried out in which the anisotropic sphere was allowed to rotate with a rotational diffusion constant of $k_BT/8\pi\eta a^3$. For more details regarding these prototypical studies, the reader is referred to reference (23). It can be seen that the simulations are in excellent agreement with analytic results where the latter are available. Note the large increase in k in the presence of screened or unscreened Coulombic attraction arising between two oppositely charged ions of elementary charge magnetude in a dielectric medium like water ($\varepsilon = 78$). The inclusion of HI decreases k by 30% in the no-force case but only by 6% with attractive forces present. In the case of the hemisphere model, the inclusion of

Table I. Two Reactive Spheres

Reactivity[i]	Hydrodynamic Interaction	Rotation	Interparticle Forces[ii]	b(Å)	q(Å)	K(simulation) (iii)	K(analytic)
uniform	none	-	none	3	8	1.03 ± 0.07	1.00
uniform	none	-	$-\hat{r}Q^2/\varepsilon r^2$	5	10	7.18 ± 0.14	7.31
uniform	none	-	$-\hat{r}Q^2 e^{-\kappa r}/\varepsilon r^2$	3	8	4.81 ± 0.12	4.80
uniform	included	-	none	3	8	0.72 ± 0.06	0.72
uniform	included	-	$-\hat{r}Q^2/\varepsilon r^2$	5	10	7.08 ± 0.14	6.82
uniform	included	-	$-\hat{r}Q^2 e^{-\kappa r}/\varepsilon r^2$	5	10	4.66 ± 0.14	4.40
hemisphere	none	none	none	5	7	.711 ± 0.032	~.70(iv)
hemisphere	none	none	none	30	60	.700 ± 0.041	~.70(iv)
hemisphere	none	included	none	3	5	.785 ± 0.012	~.80(iv)
hemisphere	none	none	$-\hat{r}Q^2/\varepsilon r^2$	5	7	7.29 ± 0.14	-
hemisphere	none	included	$-\hat{r}Q^2/\varepsilon r^2$	5	10	7.39 ± 0.16	-
hemisphere	none	none	$-\hat{r}Q^2 e^{-\kappa r}/\varepsilon r^2$	5	10	4.92 ± 0.15	-
hemisphere	none	included	$-\hat{r}Q^2 e^{-\kappa r}/\varepsilon r^2$	5	10	4.97 ± 0.18	-

(i) uniform - both spheres uniformly reactive; hemisphere - one sphere uniformly reactive but only half the surface of remaining sphere reactive.
(ii) Q = 1 elementary charge unit, ε = 78 = solvent dielectric constant, κ = 0.1 $Å^{-1}$ (corresponding to $[Na^+]$ = 0.1 M).
(iii) $K = k/k_D°$ where $k_D° = 4\pi(a_1 + a_2)D_{rel}$.
(iv) From Figures II and III of reference 10.

attractive Coulombic forces has an even more dramatic effect on k, restoring it to the value observed when both spheres are uniformly reactive. Evidently, attractive forces serve to hold the spheres together long enough for them to achieve a favorable configuration for reaction. This is likely to be a feature of some enzyme-substrate interactions, Since surface charges on enzymes are ubiquitous. The effect of rotation is more modest, having the largest effect in the absence of direct attractive forces. Since the rotational diffusion constant of a particle varies roughly as a^3, the effect of rotation on reaction rate is expected to be small when an anisotropic target (enzyme) is much larger than the substrate.

Sphere and Dumbell Dimer. Dumbell dimers reacting with a spherical target represent the simplest case of structured reactants. The model used is depicted in Figure 2. The radii of the target (subunit 1) and dimer (subunits 2 and 3) were 2.0 and 0.5 Å, respectively. The target sphere and either one or both dimer subunits were taken to be reactive. The criteria for a reactive collision were $R \leq 3$ Å, and (for the cases with only one dimer subunit reactive) $\Theta < 90°$. To study the effects of direct forces on reaction rates, variable charges (Q_i) were placed at the centers of the subunits.

To determine a rate constant, $k_D(b)$ and the recombination probability β must be obtained. For particles interacting via a centrosymmetric potential of mean force, Eq. 2 can be used to obtain $k_D(b)$. For most of the model studies considered in this section, however, Eq. 2 is not strictly valid since the potential of mean force has an angular (Θ) dependence. Making the reasonable assumption that the relative orientation of the dimer follows a Boltzmann distribution at R = b, an equation similar to Eq. 2 can be derived (34). To determine β, dynamical trajectories are computed using Eq. (5) or equations derived therefrom, (34) starting at R = b = 8 Å and with relative orientations selected at random from a Boltzmann distribution. Trajectories were terminated at $R \geq q = 10$ Å and β was determined from the results of 5,000 to 10,000 separate trajectories. The interested reader is referred to reference 34 for more details. The rate constant results are summarized in Table II.

To elucidate more directly the role of electrostatic and hydrodynamic forces in "steering" the dimer toward productive collision geometries, the distribution of relative orientations f(R, cosΘ) as a function of R was determined from the simulations. Specifically, f(R, cosΘ) represents the probability that a reactive/unreactive dimer at R has an orientation lying between cos$\Theta \pm 0.1$. In the special case of an isotropic distribution, f(R, cosΘ) = 0.1 since Θ-space has been divided into ten equivalent "bins" ($-1 \leq \cos\Theta \leq +1$). The observed distribution is nearly identical to the Boltzmann distribution for R > 4.5 (34). For R close to the reaction radius of 3 Å, however, this is not the case. Three examples are shown in Figures 3-5 where R = 3.05 ± 0.05Å. A characteristic feature of dimers which approach this close without eventually reacting (unfilled symbols) is that they are in an unfavorable orientation. The fact that dimers in reactive as well as unreactive trajectories have high probability of being in an unfavorable orientation is interpreted in the following way. A dimer which approaches the target closely in a favorable orientation tends to react quickly

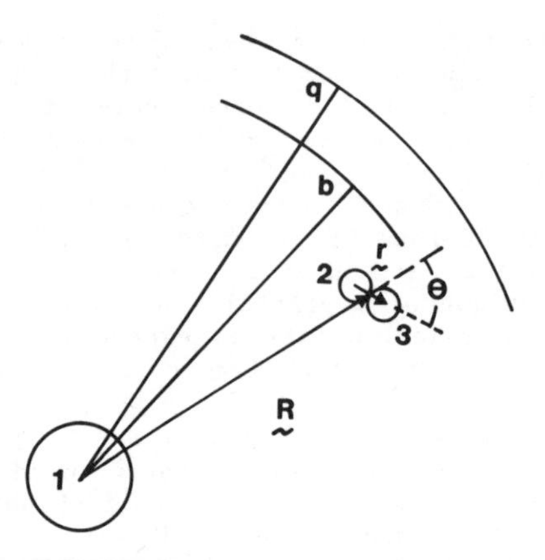

Fig. 2. Sphere - Dimer Model. The target sphere (subunit 1) and one or both dimer subunits (2 and 3), are uniformly reactive. The target sphere radius is 2 Å and the touching spheres of the dimer each have a radius of 0.5 Å. Charges (Q_1, Q_2, Q_3) are placed at the subunit centers.

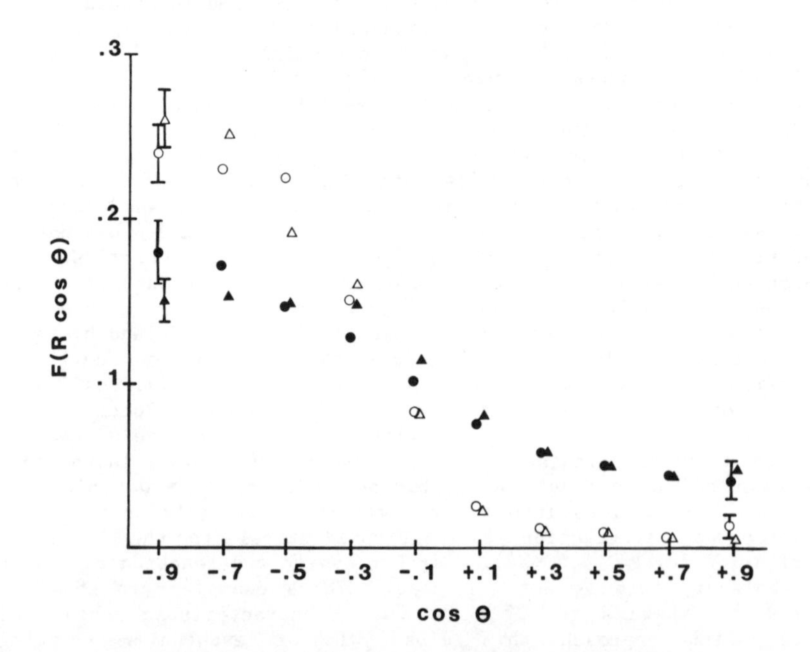

Fig. 3. Orientation Factor Near the Reactive Surface. R = 3.05 ± .05 Å, $Q_1 = Q_2 = Q_3 = 0$. Only subunit 2 of the dimer is reactive. Filled/empty circles (●/o) represent reactive/unreactive trajectories where hydrodynamic interaction (HI) is included, and filled/empty triangles (▲/Δ) represent reactive/unreactive trajectories where HI is not included. Error bars are placed on only certain data points.

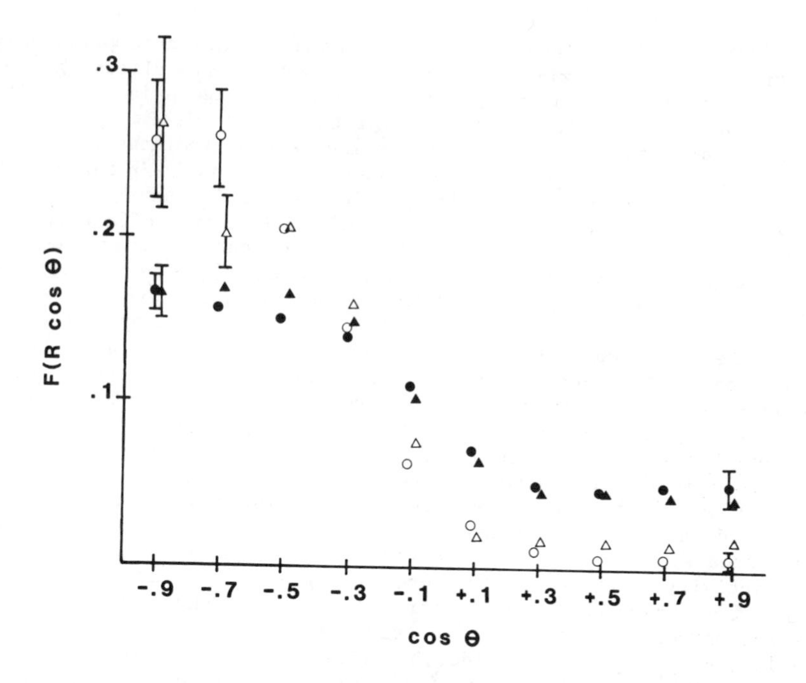

Fig. 4. Orientation Factor Near the Reactive Surface. Same as Fig. 3, but $Q_1 = e$, $Q_2 = -e$, $Q_3 = 0$.

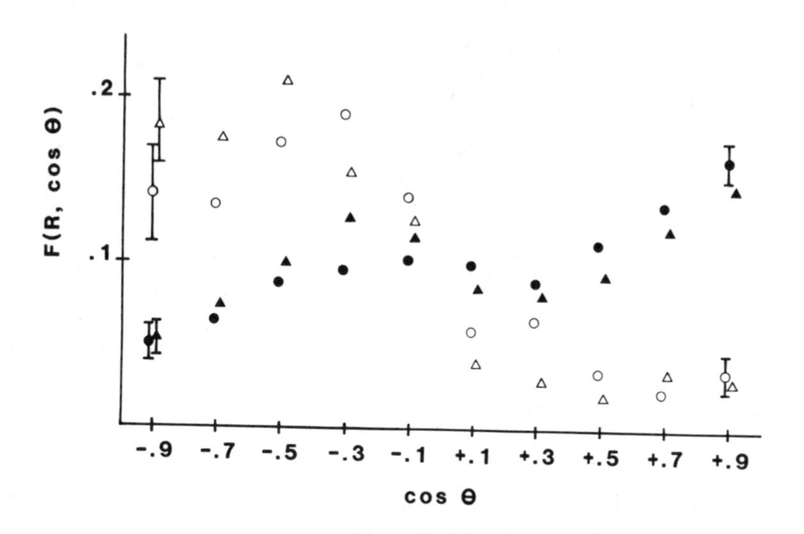

Fig. 5. Orientation Factor Near the Reactive Surface. Same as Figs. 3 and 4, but $Q_1 = 2e$, $Q_2 = -e$, $Q_3 = +e$. Unlike Fig. 3 (no charges) or Fig. 4 (net Coulomb charges on both reactants), the dimer in this case is a pure electric dipole.

whereas a dimer in an unfavorable orientation spends a comparatively long time in close proximity to the target eventually diffusing away (unreactive) or reorienting to a favorable configuration at which point it quickly reacts. This interpretation is also supported by the rate constants which do not show a substantial increase when both dimer subunits are made reactive. Net charges on both target and dimer and to a lesser extent hydrodynamic interaction have a substantial effect on overall rate but a surprisingly small effect on the dimer orientations. E.g., Figure 3 and 4 are nearly superimposable despite the > 2-fold difference in rate. Comparing Figs. 3 and 4 to Figure 5, it is seen that dipolar forces have a substantial effect on orienting the dimer and, from Table II, also a significant effect on overall rate. Comparing the third and sixth lines of Table II it would appear that attraction of the dipole by the inhomogeneous electric field of the target makes a smaller contribution to the rate enhancement than does the orientational steering effect.

From the sphere-dimer studies, two major conclusions emerge. The first is that the trajectory method can be extended to structured reactants with anisotropic reactivity and anisotropic direct forces and hydrodynamic interactions. The second major conclusion is that complicated electrostatic interactions between species with anisotropic reactivity can "steer" the approaching particles into favorable orientations and enhance the reaction rate. For these model studies, rate enhancements up to 20% have been obtained. The second conclusion is likely to be of considerable relevance to molecular biology. In the third and final series of simulations, the Brownian dynamics trajectory method is applied to a particular biological system.

Superoxide Dismutase and Superoxide. Electrostatic interactions influence the rates of many biomolecular associations (15). A particular example of this is the diffusion controlled dismutation of superoxide (O_2^-) catalyzed by the enzyme copper, zinc superoxide dismutase (SOD) (35-36). Although both substrate and enzyme are negatively charged at physiological pH, the reaction rate is high and increases with decreasing ionic strength at moderate salt concentrations (35). This is opposite of the trend expected on the basis of the net charges and may be due to local electrostatic interactions which may serve to steer O_2^- into the active site of SOD.

For the initial studies described here, the SOD dimer, which is the active form of the enzyme, was modeled as a sphere of 30 Å radius. Two reactive patches corresponding to the active site regions of the dimer were defined by the surface area lying within 10° of an axis running through the center of the sphere (Figure 6). Five charges were embedded within the sphere to reproduce the monopole, dipole, and quadrupole terms associated with the charged groups in the 2 Å crystallographic structure of bovine erythrocyte SOD (37) available through the Protein Data Bank (38). The net charge is -4 in units of the protonic charge and the dipole moment approximately vanishes due to the symmetry of the dimer. A dielectric constant of 78 was assumed throughout the system. On the basis of previous experimental (39) and computational (36) studies, this should provide a reasonable as well as simple description of the direction and magnitude of electrostatic forces on O_2^-. The O_2^- molecule was represented by a sphere of radius 1.5 Å with a central charge of -1.

Table II. Relative Rate Constants for Sphere - Dimer Models

Q_1	Q_2	Q_3	hydrodynamic interaction	k/k_o [(a)]
0	0	0	none	.906 ± .018
0	0	0	included	.740 ± .008
0	0	0	none	1.068 ± .017[(b)]
1	-1	1	none	1.039 ± 0.18
1	-1	1	included	.802 ± .019
1	-1	1	none	1.069 ± 0.016[(b)]
2	-1	1	none	1.080 ± .020
2	-1	1	included	.873 ± .017
1	-1	0	none	2.51 ± .05
1	-1	0	included	2.21 ± .05

(a) $k_o = 4\pi D_M r_o$ where $r_o = 3$ Å and D_M is the effective relative diffusion constant (see reference 34).

(b) Both subunits of dimer (2 and 3) reactive. In all other simulations, only subunit 2 is reactive.

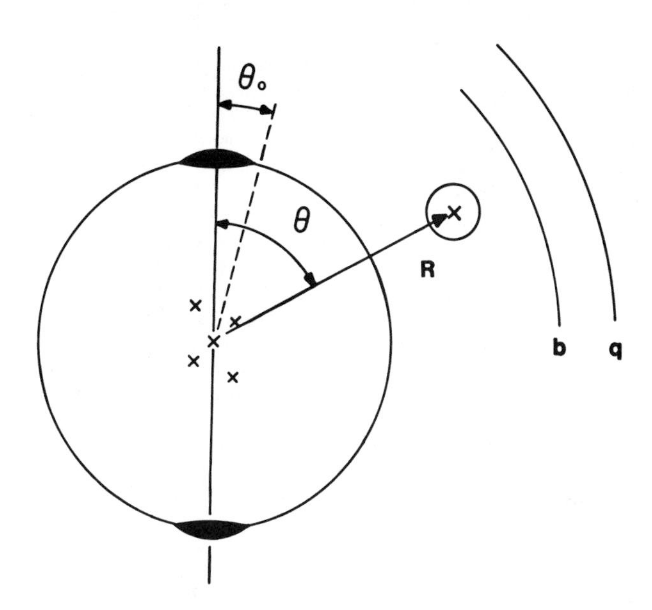

Fig. 6. Model of SOD - Superoxide. Crosses (X) indicate positions of charges. Active sites are indicated by the dark caps on the SOD sphere; $\Theta_o = 10°$.

Hydrodynamic interaction between SOD and O_2^- was ignored in this preliminary study. Between 25,000 and 100,000 trajectories were carried out in each simulation with typical b and q values of 300 and 500 Å respectively. For additional details, see References 40 and 41.

Table III summarizes some calculations carried out to explore what effects contribute to the high reactivity of SOD. For the native-like model with a monopole charge of -4, inclusion of the (non-centrosymmetric) quadrupole increases the reaction rate by 40%. The quadrupole evidently helps to steer O_2^- into the active site. Parallel simulations were also carried out in which monopole charges of 0 and +4 were used. Although increasing the monopole charge from -4 to 0 to +4 increased the rates by factors of 2.5 and 5, respectively, the steering effect is present in each case. This suggests that the enhancement in rate due to steering by local electrostatic interactions will persist in the presence of added salt, which will suppress the effects of the monopole field more strongly than those of the shorter-ranged quadrupole field.

The qualitative effect of added salt has been examined directly by simulations using a screened potential of the Debye-Hückel type and the results are shown in Figure 7. Above an ionic strength of about 3×10^{-2} M, the rate decreases with added salt as the steering field due to the quadrupole is screened. A qualitatively similar trend is observed experimentally. In future work, we plan to examine increasingly detailed models that include the irregular surface topography and full charge distribution of SOD, individual salt ions, local dielectric constant variations, etc.

Summary and Conclusions

A principal aim of this work was to demonstrate the utility and generality of a new simulation method for determination of the rates and mechanisms of diffusion controlled reactions. With regard to the role of electrostatic interactions in diffusion controlled reactions, several conclusions can be made on the basis of the model studies discussed in the previous section. _Net_ charges have a significant and, in some cases, dramatic effect on overall rate but may not play an important role in steering the reactive species into orientations favorable for reaction. Net charges do play an important role in bringing the species together. Local electrostatic forces can help to steer species into productive orientations. This was manifest in rate enhancement of 20 to 60% in the above model studies; it is likely that larger effects occur in other systems. Such steering effects are likely to be important in molecular biology since virtually all biomolecules have complex charge distributions on their surfaces. Furthermore, the relative importance of local electrostatic interactions is expected to be greatest at moderate (physiological) salt concentrations. Hydrodynamic interactions were observed to decrease the rate by 5 to 30% but had little effect on "steering".

In future work, the methods illustrated in this paper will be applied to a variety of problems in macromolecular kinetics. More detailed studies of substrate binding to superoxide dismutase and antigen binding to antibody molecules are in progress. Other studies that are planned or in progress include the examination of Coulombic contributions to polymer growth and to DNA-ligand interactions.

Table III. Relative Rate Constants for Various SOD Models

SOD Charge	Model	k/k_o*
−4	Monopole	0.056 ± .003
−4	Monopole plus quadrupole	0.079 ± .004
0	Monopole	0.12 ± .01
0	Monopole plus quadrupole	0.19 ± .01
+4	Monopole	0.26 ± .02
+4	Monopole plus quadrupole	0.41 ± .03

*The rate constant k is normalized by that, k_o, expected for an SOD model with no embedded charges and a uniformly reactive surface. At least 25,000 trajectories were computed for each model.

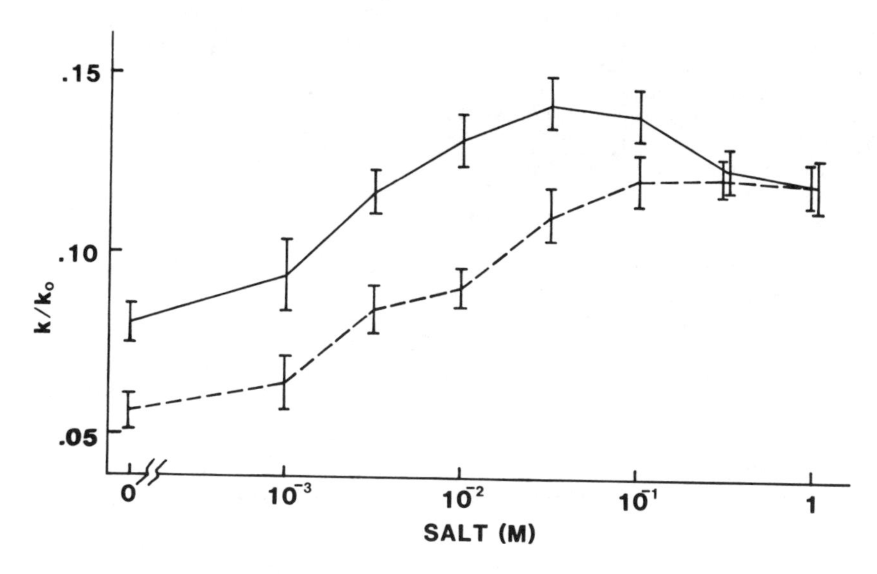

Fig. 7. Dependence of Relative Rate on Ionic Strength ("Salt"). Solid and dotted lines connect monopole + quadrupole and monopole rates, respectively. The electrostatic potential energy between the charge on O_2^- (Q_1) and a particular SOD charge Q_2 separated by r was taken to be $Q_1Q_2e^{-\kappa r}/\varepsilon r$ where ε is the dielectric constant (=78) and κ is the Debye-Hückel parameter.

Acknowledgments

This work was supported in part by grants from the Robert A. Welch Foundation and NIH (Houston), the Research Corporation and the Petroleum Research Fund as administered by ACS (Tennessee Tech). SAA is the recipient of an NSF Presidential Young Investigator Award and a Dreyfus Grant for Young Faculty in Chemistry. SHN is the recipient of an NIH Career Development Award. JAM is an Alfred P. Sloan Fellow and is the recipient of NIH Career Development and Dreyfus Teacher-Scholar Awards.

Literature Cited

1. Smoluchowski, M.V. Phys. Z. 1916, 17, 557.
2. Debye, P. Trans. Electrochem. Soc. 1942, 82, 265.
3. Friedman, H.L. J. Phys. Chem. 1966, 70, 3931.
4. Deutch, J.M.; Felderhof, B.U. J. Chem. Phys. 1973, 59, 1669.
5. Northrup, S.H.; Hynes, J.T. J. Chem. Phys. 1979, 71, 871.
6. Calef, D.F.; Deutch, J.M. Annu. Rev. Phys. Chem. 1983, 34, 493.
7. Solc, K.; Stockmayer, W.H. Int. J. Chem. Kinet. 1973, 5, 733.
8. Schurr, J.M.; Schmitz, K.S. J. Phys. Chem. 1976, 80, 1934.
9. Samson, R.; Deutch, J.M. J. Chem. Phys. 1978, 68, 285.
10. Shoup, D.; Lipari, G.; Szabo, A. Biophys. J. 1981, 36, 697.
11. McCammon, J.A.; Northrup, S.H. Nature 1981, 293, 316.
12. Szabo, A.; Shoup, D.; Northrup, S.H.; McCammon, J.A. J. Chem. Phys. 1982, 77, 4484.
13. McCammon, J.A. Rep. Progr. Physics 1984, 47, 1.
14. van Leeuwen, J.W. FEBS Lett. 1983, 156, 262.
15. Neumann, E. In "Structural and Functional Aspects of Enzyme Catalysis"; H. Eggerer; R. Huber, Eds.; Springer: Berlin, 1981; pp. 45-58.
16. Noyes, R.M. Prog. Reac. Kin. 1961, 1, 128.
17. Wilemski, G.; Fixman, M. J. Chem. Phys. 1973, 58, 4009.
18. Wilemski, G.; Fixman, M. J. Chem. Phys. 1974, 60, 866; 878.
19. Doi, M. Chem. Phys. 1975, 11, 107, 115.
20. James, C.; Evans, G.T. J. Chem. Phys. 1982, 76, 2680.
21. Berg, O.G. (preprint entitled "Diffusion-Controlled Protein-DNA Association").
22. Zientra, G.P.; Nagy, J.A.; Freed, J.H. J. Chem. Phys. 1980, 73, 5092.
23. Northrup, S.H.; Allison, S.A.; McCammon, J.A. J. Chem. Phys. 1984, 80, 1517.
24. Ermak, D.L.; McCammon, J.A. J. Chem. Phys. 1978, 69, 1352.
25. Fixman, M. J. Chem. Phys. 1978, 69, 1527; 1538.
26. Fixman, M. (preprint entitled "Implicit Algorithm for Brownian Dynamics of Polymers").
27. Pear, M.R.; Weiner, J.H. J. Chem. Phys. 1979, 71, 212.
28. Lamm, G.; Schulten, K. J. Chem. Phys. 1983, 78, 2713.
29. Lamm, G. (preprint entitled "Extended Brownian Dynamics. III.").
30. "IMSL Library 7 Reference Manual," IMSL International and Statistical Libraries, Inc., Houston, Texas 1975.

31. Wolynes, P.G.; Deutch, J.M. J. Chem. Phys. 1976, 65, 450, 2030.
32. Ryckaert, J.P.; Ciccotti, G.; Berendsen, H. J. Comp. Phys. 1977, 23, 327.
33. Allison, S.A.; McCammon, J.A. Biopolymers 1984, 23, 167; 363.
34. Allison, S.A.; Srinivasan, N.; McCammon, J.A.; Northrup, S.H. J. Phys. Chem. 1984, 88, 6152.
35. Cudd, A.; Fridovich, I. J. Biol. Chem. 1982, 257, 11443.
36. Getzoff, E.D.; Tainer, J.A.; Weiner, P.K.; Kollman, P.A.; Richardson, J.S.; Richardson, D.C. Nature 1983, 306, 287.
37. Tainer, J.A.; Getzoff, E.D.; Beem, K.M.; Richardson, J.S.; Richardson, D.C. J. Mol. Biol. 1982, 160, 181.
38. Bernstein, F.C. et al. J. Mol. Biol. 1977, 112, 535.
39. Rees, D.C. J. Mol. Biol. 1980, 141, 323.
40. Allison, S.A.; McCammon, J.A. J. Phys. Chem. 1985, 89, 1072.
41. Allison, S.A; Ganti, G.; McCammon, J.A. J. Phys. Chem. (in press).
42. Allison, S.A.; Northrup, S.H.; McCammon, J.A. J. Chem. Phys. (in press).

RECEIVED June 10, 1985

19

Interaction of Anionic Detergents with Cationic Residues in Polypeptides

Conformational Changes

Wayne L. Mattice

Department of Chemistry, Louisiana State University, Baton Rouge, LA 70803

For some time it has been known that sodium dodecyl sulfate exerts a strong ordering effect on cationic homopolypeptides. In contrast, the anionic detergent exerts a negligible effect on the thermally-induced order-disorder transitions seen in the poly(hydroxyalkyl-L-glutamine) series of water-soluble nonionic homo- and copolypeptides. The change in the helix propagation parameter, s, is on the order of 0.01 when the hydroxyamyl-L-glutaminyl residue is transferred from water to 0.003 M sodium dodecyl sulfate. Circular dichroism spectra obtained with several homologous peptide hormones in sodium dodecyl sulfate can be rationalized by reasonable extension of the results obtained with synthetic polypeptides. The necessary change in the helix propagation parameter for cationic residues is more than two orders of magnitude greater than that seen with nonionic hydroxyalkyl-L-glutaminyl residues.

Successful prediction of the conformational properties of a protein from its amino acid sequence is one of the major tasks of macromolecular physical chemistry. When the protein is completely disordered, the major difficulty is the proper averaging over all accessible conformations. In the special case where the random coil is unperturbed by long-range interactions, matrix methods are ideally suited for the averaging (1). For certain disordered proteins, the agreement between experimental and theoretical dimensions is quite good, as shown by the two examples in Table I. Experimental mean square radii of gyration are those measured for myelin basic protein by Krigbaum and Hsu (2) and for crosslinked tropomyosin by Holtzer et al. (3). The values reported in Table I are corrected for the expansion produced by the excluded volume effect (2,4). This correction is negligible in the case of myelin basic protein (2). There are no adjustments of any parameters used in the computation of the theoretical values (4,5). Agreement between experiment and theory is quite good.

0097-6156/86/0302-0232$06.00/0

Table I. Unperturbed Root-mean-square Radii of Gyration (nm) for Two Disordered Proteins

Protein	Solvent	Exp.	Theory	Diff (%)
Myelin basic protein	Water	4.56	4.64	2
Tropomyosin dimer (crosslinked)	5 M Guanidine HCl	8.33	8.75	5

The necessity for a proper averaging over a multitude of conformations remains when the protein is partially ordered. The power of the matrix methods can still be utilized if the dominant ordering interactions are of relatively short range, as they are in α helix formation (6). The conformational partition function, Z, for a chain of n residues is formulated as shown in Equation 1.

$$Z = J^* U_1 U_2 \dots U_n J \qquad (1)$$

Each statistical weight matrix, U, contains elements weighting disordered residues and residues at the ends and in the interior of helical segments. The pertinent statistical weights can, in principle, be extracted from studies of homopolypeptides in aqueous solution. For a variety of technical reasons, this direct approach is not available with most amino acid residues of biological interest. The desired statistical weights are obtained instead by a more elaborate procedure, in which a poly(hydroxyalkyl-L-glutamine) serves as the "host" in a "host-guest" copolypeptide (7). Using this approach, Scheraga and coworkers have determined σs and s (the statistical weights for helix initiation and propagation, respectively) in water for 18 of the 20 amino acid residues commonly found in proteins (8). This set of σ and s permits computation of a helical content as

$$f = n^{-1} \sum_{i=1}^{n} (\partial \ln Z / \partial \ln s_i)_{\sigma,n,s_j \neq s_i} \qquad (2)$$

In general, the helical content calculated in this manner is in good agreement with that deduced from circular dichroism spectra only when long-range interactions are inconsequential, as is the case with tropomyosin at temperatures high enough so that the dimer has been completely dissociated (9).

Several peptides and proteins that play important roles in the nervous system are predominantly disordered in dilute aqueous solution, but develop order upon addition of anionic lipids. The conformational changes detected by circular dichroism can often be rationalized using the matrix formulation for Z, but with altered statistical weights for a few crucial residues. Statistical weights obtained directly from Scheraga's "host-guest" studies in water (8), or estimated from such studies, are used for most of the amino acid residues. The exceptions are increases in the helix initiation and propagation parameters for cationic arginyl, histidyl, and lysyl residues when they are in an aqueous environment containing lipid or anionic detergent. The increase in the propagation parameters, s, is from 0.69-1.03 (8) in water to about 1.7 (10,11).

The rationale for the changes in σ and s is supplied by studies of simpler model systems. Cationic homopolymers of Arg, Lys, and His are disordered in water, but they readily adopt ordered structures in the presence of sodium dodecyl sulfate. The conformation of cationic poly(L-arginine) changes from a random coil to an α helix upon the addition of small amounts of sodium dodecyl sulfate (12), as shown in Figure 1. Yang (13) has shown that lysine-containing copolymers in sodium dodecyl sulfate favor the β sheet over the α helix when the sequence is $(Lys\text{-}Xxx)_n$, where Xxx may be Lys or some other residue. For other sequences (including those more typical of naturally occurring peptides) Lys favors the α helix over the β sheet in the presence of dodecyl sulfate. Circular dichroism spectra demonstrate that sodium dodecyl sulfate induces a conformational transition in poly(L-histidine), but the ordered structure formed is not easily identified due to contributions to the spectra from electronic transitions in the imidazole group (12).

Sodium dodecyl sulfate has no effect on helix formation by an anionic homopolypeptide such as poly(L-glutamic acid), as expected from coulombic considerations (14). The size of the effect of the anionic detergent on helix formation by nonionic homopolypeptides has been deduced from studies of members of the poly(hydroxyalkyl-L-glutamine) series. These polypeptides are among the few nonionic polypeptides for which helix-coil transitions can be measured in water. Sodium dodecyl sulfate exerts only a very small effect on the circular dichroism of copolymers of hydroxypropyl-L-glutamine and hydroxyamyl-L-glutamine (15), as shown in Figure 1. The small effect is in marked contrast to the dramatic changes seen upon interaction of a cationic homopolypeptide, such as poly(L-arginine), with the oppositely charged detergent. In Figure 1, the concentration of sodium dodecyl sulfate in the solution used for spectrum HG2 is 30 times larger than the concentration in the solution used for spectrum A2. Upon comparison of these spectra with those obtained in the absence of sodium dodecyl sulfate (HG1 and A1), it is evident that the conformation of the nonionic copolypeptide is much less sensitive to the anionic detergent than is the conformation of cationic poly(L-arginine). Analysis of spectra obtained with hydroxypropyl-L-glutamine - hydroxyamyl-L-glutamine copolypeptides with other compositions shows that a 0.003 M solution of sodium dodecyl sulfate increases s by about 0.003 for hydroxypropyl-L-glutaminyl residues and 0.012 for hydroxyamyl-L-glutaminyl residues (15). Circular dichroism spectra of the related hormonal peptides vasoactive intestinal peptide, secretin, and glucagon suggest that the anionic detergent produces an increase in s for cationic residues of about 0.7-1.2 (11). The effects on the cationic residues are at least two orders of magnitude larger than those seen with the hydroxyalkyl-L-glutamines. Clearly the most important helix-producing interaction of the anionic detergent is exerted at sites occupied by cationic residues.

If the pertinent σ and s are available, a helix probability profile, depicting the probability for a helical placement as a function of the position in the chain, can be extracted from Z.

$$f_i = (\partial \ln Z / \partial \ln s_i)_{\sigma, n, s_j \neq s_i} \qquad (3)$$

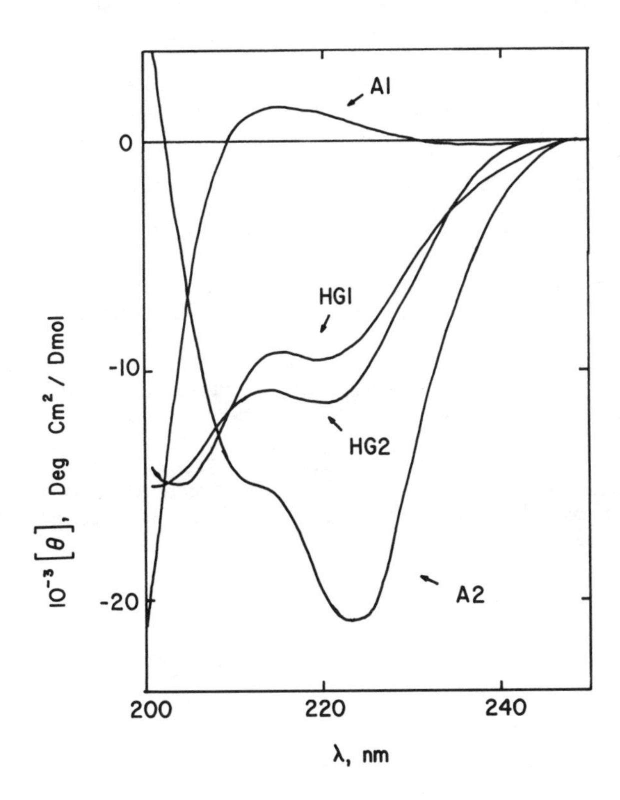

Figure 1. Circular dichroism spectra at 24-25° C for cationic poly(L-arginine) and a nonionic random copolypeptide of hydroxypropyl-L-glutamine (83%) and hydroxyamyl-L-glutamine (17%). (A1) Poly(L-arginine) in water, (A2) poly(L-arginine) in 0.0001 M sodium doecyl sulfate, (HG1) hydroxyalkylglutamine copolypeptide in water, and (HG2) hydroxyalkylglutamine copolypeptide in 0.003 M sodium dodecyl sulfate. Spectra for poly(L-arginine) are from reference 12, and spectra for the random copolypeptide are from reference 15.

This information is used to predict probable locations for helices. The approach can be illustrated by a consideration of the hormonal peptide glucagon (11). Its sequence is depicted in Figure 2. The combination of circular dichroism spectroscopy and statistical mechanics indicates little helical content in water (dashed line in Figure 2). There is, however, an enhancement in helicity, particularly at residues 13-24, in the presence of 0.003 M sodium doecyl sulfate (solid line). Wuthrich's combination of 500 MHz NOESY experiments and distance geometry calculations finds a helical segment at residues 17-29 ("#" in Figure 2) in micelle-bound glucagon (16). Both approaches focus attention on a helical segment containing about three turns, and they agree on the location of two of the three turns. The helical content is enhanced primarily in that portion of the chain that has the highest helix-forming potential when the environment is simply water. Plausible adjustment in our parameters would not move the helical segment to residues 17-29 ("#" in Figure 2). Calculations based on Equation 1 routinely show that residues located at the extreme ends of a chain have an overwhelming preference for the random coil state when o and s are assigned realistic values.

The basis for assigning s = 1.7 and σ = 0.05 for Arg, His, and Lys can be traced to our investigation of several proteins in 1976 (10). At that time, σ and s for these three residues were treated as adjustable parameters. We assumed (1) all three residues had the same s, (2) all three residues had the same σ, and (3) an empirical correlation related σ and s. This set of assumptions reduced the number of independently adjustable parameters from six

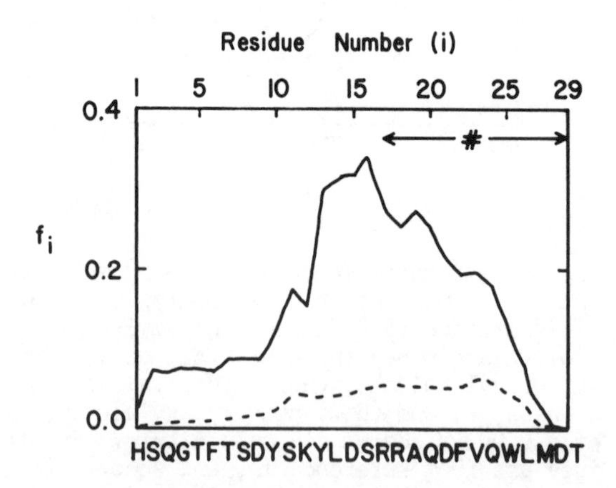

Figure 2. Helix probability profiles deduced for glucagon in water (dashed line) and in aqueous sodium dodecyl sulfate (solid line). Along the bottom of the Figure is the amino acid sequence of glucagon, using A = Ala, D = Asp, F = Phe, G = Gly, H = His, K = Lys, L = Leu, M = Met, Q = Gln, R = Arg, S = Ser, T = Thr, V = Val, W = Trp, Y = Tyr. The text describes the significance of "#". The helix probability profiles are from reference 11.

to one. The assumptions cannot be correct in detail. Nevertheless, they have proven useful in the analysis of circular dichroism spectra, as exemplfied by the glucagon study described above. Improved reliability of the computed helix probability profiles must await the determination of σ and s for the Arg, His, and Lys residues separately.

Acknowledgment

This work was supported by National Science Foundation research grant PCM 81-18197.

Literature Cited

1. Flory, P. J. Macromolecules 1974, 7, 381-92.
2. Krigbaum, W. R.; Hsu, T. S. Biochemistry 1975, 14, 2542-6.
3. Holtzer, A.; Clark, R.; Lowey, S. Biochemistry 1965, 4, 2401-11.
4. Mattice, W. L. Macromolecules 1977, 10, 516-20.
5. Mattice, W. L.; Robinson, R. M. Biopolymers 1981, 20, 1421-34.
6. Poland, D.; Scheraga, H. A. "Theory of Helix-Coil Transitions in Biopolymers"; Academic Press: New York, 1970.
7. von Dreele, P. H.; Lotan, N.; Ananthanarayanan, V. S.; Andreatta, R. H.; Poland, D.; Scheraga, H. A. Macromolecules 1971, 4, 408-17.
8. Sueki, M.; Lee, S.; Powers, S. P.; Denton, J. B.; Konishi, Y.; Scheraga, H. A. Macromolecules 1984, 17, 148-55.
9. Mattice, W. L.; Srinivasan, G.; Santiago, G. Macromolecules 1980, 13, 1254-60. Holtzer, M. E.; Holtzer, A.; Skolnick, J. Macromolecules 1983, 16, 173-80.
10. Mattice, W. L.; Riser, J. M.; Clark, D. S. Biochemistry 1976, 15, 4264-72.
11. Robinson, R. M.; Blakeney, E. W.; Mattice, W. L. Biopolymers 1982, 21, 1217-28.
12. McCord, R. W.; Blakeyen, E. W.; Mattice, W. L. Biopolymers 1977, 16, 1319-29.
13. Kubota, S.; Ikeda, K.; Yang, J. T. Biopolymers 1983, 22, 2219-36, 2237-52.
14. Fasman, G. D.; Lindblow, C.; Bodenheimer, E. Biochemistry 1964, 3, 155-66.
15. Overgaard, T.; Erie, D.; Darsey, J. A.; Mattice, W. L. Biopolymers 1984, 23, 1595-603.
16. Braun, W.; Wider, G.; Wuthrich, K. J. Mol. Biol. 1983, 169, 921-48.

RECEIVED July 3, 1985

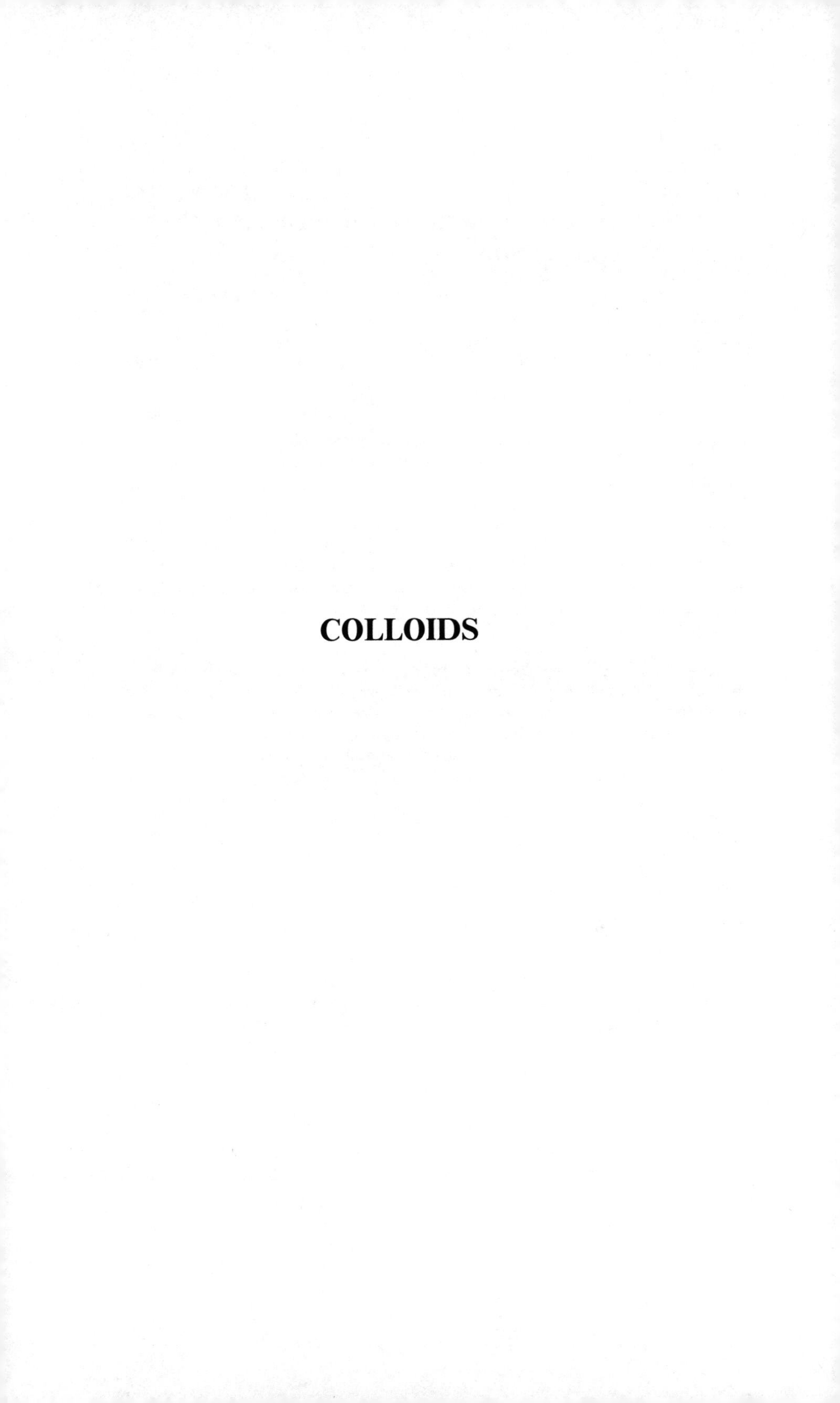

COLLOIDS

20

Complex Coacervation of Acid-Precursor Gelatin with a Polyphosphate

T. Lenk and C. Thies

School of Engineering and Applied Science, Washington University, St. Louis, MO 63130

Complex coacervation is a phenomenon by which an aqueous solution of oppositely charged polyelectrolytes separates into two distinct phases. The more dense phase is called the complex coacervate or coacervate. It is a relatively concentrated polyelectrolyte solution. The second phase, a relatively dilute polyelectrolyte solution, is called the equiibrium liquid. The difference in concentration of the coacervate and equilibrium liquid phases is determined by the intensity of the coacervation interaction. The more intense this interaction is, the greater the concentration difference.

Brungenberg de Jong and coworkers carried out the first extensive studies of complex coacervation (1). They characterized the gelatin-gum arabic coacervation system, a system that later was developed into a process capable of producing microcapsules loaded with a variety of lyophobic materials (2). More recently, an encapsulation process based on the coacervation of gelatin with a polyphosphate has been reported (3). The present paper describes results of a study designed to characterize the gelatin-polyphosphate coacervation interaction and define how various experimental paramenters affect it.

Experimental

Materials: Table I contains ash values determined by Galbraith Laboratories, Knoxville, Tn., for the three acid precursor gelatin samples used. All were generously supplied by the Hormel Corporation, Austin, Mn., and were used as received. The polyphosphate was sodium hexametaphosphate (Calgon Condition 206, Calgon Division of Merck, Pittsburgh, Pennsylvania). The pH of all samples was adjusted with reagent grade acetic acid.

Coacervation Procedure: The coacervation procedure involved weighing a 10 wt. percent gelatin solution at 55°C into a capped graduated glass centrifuge tube (15 ml capacity). The desired amount of polyphosphate solution was added as a 5 wt. percent solution. After mixing and equilibration at 55°C for 30 minutes,

0097-6156/86/0302-0240$06.00/0

the pH was adjusted with acetic acid and deionized water was added to give a total sample volume of 14 ml. The ionic strength of the coacervation varied somewhat with the pH because varying amounts of acetic acid were used to adjust pH and no supporting electrolyte was present. The system was mixed and allowed to equilibrate 40 to 60 minutes at 55°C. During this equilibration, the coacervate and equilibrium liquid separated into two well-defined layers. The volume of each layer was recorded. A 5 ml aliquot of equilibrium liquid was removed from each sample tube and weighed immediately into a tared aluminum weighing pan. This aliquot was dried to constant weight at 110°C to give the total solids content of the equilibrium phase. Another aliquot of equilibrium liquid (1 ml) was used to determine the phosphate content of this phase. A spectrophotometric (820 nm) assay procedure based upon the complexation of phosphate with an aqueous sulfuric acid ammonium molybdate-ascorbic acid mixture was used (4). In order to eliminate interference caused by gelatin, each one ml aliquot of equilibrium liquid was digested with one ml of concentrated HCl at 85°C for 25 hours before the phosphate assay was made. A calibration curve for the phosphate assay procedure was constructed from 52 standard polyphosphate and 20 standard polyphosphate-gelatin solutions treated exactly as the unknown equilibrium liquid samples. Both standard solutions fit the same calibration curve. The remaining equilibrium liquid was used to measure system pH. Preliminary experiments found no significant pH difference between the coacervate and equilibrium liquid layers. All reported solution concentrations are in weight percent.

Table I. Gelatin Characterization Data

Bloom Strength	Ash, wt. percent
150	0.50
275	0.10
300	0.24

Total solids of the equilibrium liquid, phosphate content of the equilibrium liquid, volume of coacervate, total volume of the system, and total weight of gelatin and polyphosphate in the system were used to calculate the following quantities:

1. Volume percent coacervate: (Volume of coacervate/Total volume of system) x 100.
2. Total solids content of each phase.
3. Concentration of each polymer in each phase.
4. Fraction of each polymer in the coacervate.
5. Degree of coacervation, ρ. ρ is the fraction of total polymer in the coacervate.
6. Enrichment, ε. ε is the ratio of total solids of the coacervate/total solids of the equilibrium liquid.
7. Intensity of coacervation, $\theta : \theta = \varepsilon . \rho$.
8. Gelatin/polyphosphate ratio in each phase.

All of the above quantities were plotted as a function of system pH for each complex coacervation system studied. Only a few of the many plots obtained are included in this paper. An effort was made to present results in terms of θ. In addition, several plots of total solids content versus pH are given. These graphically illustrate how a given parameter affects the pH range over which coacervation occurs as well as the total solids content of the equilibrium liquid and complex coacervate phases that form. The other data plots given illustrate specific points of interest.

The experimental procedure presented above involves a number of sequential steps that utilize relatively small volumes and weights of material. This contributed to scatter of the experimental data. Since there was no clear justification for rejecting a specific data point, all points obtained were included in the final graphs regardless of how markedly they deviated from an observed trend. The effect of data scatter on interpretation of results was minimized by assaying a number of samples for each coacervate system characterized.

Results

Figure 1 is a plot of total solids content versus pH for a 4.4 percent gelatin (275 bloom) - 0.48 percent polyphosphate mixture. The continous curve shown encloses the region in which complex coacervation occurs and two phases coexist. These two phases are a polymer-rich phase called the complex coacervate and a more dilute phase called the equilibrium liquid. The straight line that divides the curve into two parts is the total solids content of the mixture before coacervation (4.88 percent). Points that fall above this line (open circles) are total solids contents of the coacervate at various pH values. Points below this line (solid circles) are corresponding total solids contents for the equilibrium liquid.

The curve in figure 1 extends from pH 2.8 to 4.7, the pH range over which this mixture forms a complex coacervate. At lower or higher pH values, no coacervate forms and the system exists as a homogeneous solution. The solids content of the coacervate and equilibrium liquid phases converge as these pH limits are approached. Between the pH limits of coacervation, the total solids content of the coacervate phases passes through a maximum of 23 percent. This maximum coacervate solids content occurs between pH 3.5 and 3.7, the same pH region where the total solids content of the equilibrium phase passes through a minimum of 0.8 percent.

Figure 2 shows that the volume of the coacervate phase in the system described by Figure 1 passes through a minimum between pH 3.5 and 3.7. Thus, the pH region where the coacervate has maximum solids content is also the pH region where coacervate volume is a minimum. This pH range is where the strongest gelatin-polyphosphate interactions occur. The rapid fall-off in coacervate volume as the pH increases above 4 is attributed to the rapid decrease in intensity of coacervation at these pH's.

Degree of coacervation, enrichment, and coacervation intensity (θ) were calculated for each experimental point shown in figures 1 and 2. Figure 3 contains the θ values obtained as well as θ values for three other coacervation systems. All four systems have the same 9.1/1 (w/w) gelatin (275 bloom)/polyphosphate ratio. They differ only in total solids content which ranges from 7.32 down to 1.22 percent. This variation is responsible for the large

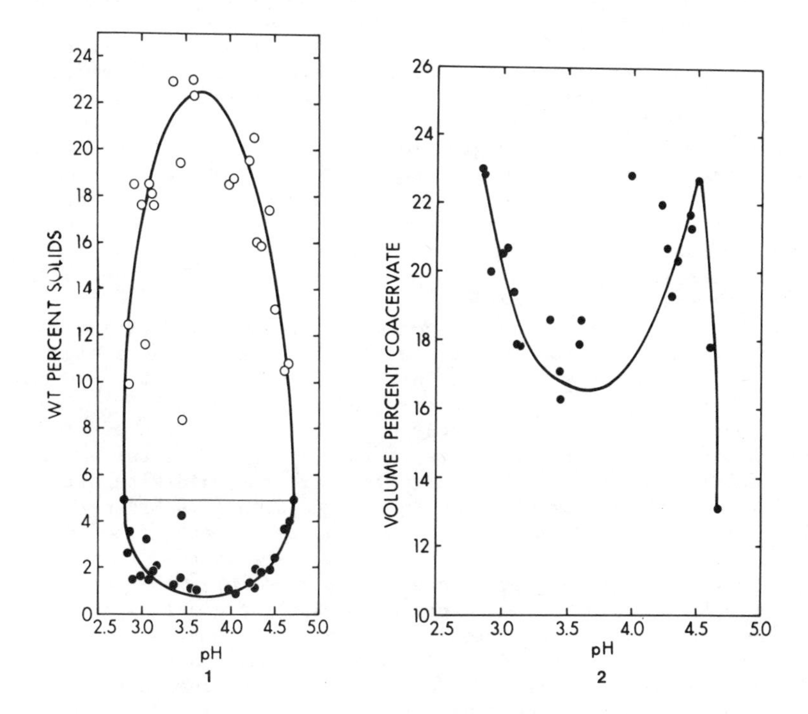

Figure 1. Plot of total solids content versus pH for a 4.4 percent gelatin (275 bloom) - 0.48 percent polyphosphate coacervation system at 55°C: 0, total solids content of the coacervate; ●, total solids content of the equilibrium liquid.

Figure 2. Plot of volume percent coacervate versus pH at 55°C for a 4.4 percent gelatin (275 bloom) - 0.48 percent polyphosphate coacervation system.

variation in θ values shown. As the coacervation system is diluted, the intensity of the coacervation reaction increases dramatically. This is particularly true between pH 3.4 and 3.9, the pH range in which θ values for all four samples pass through a maximum.

Figure 4 is another plot that graphically illustrates how initial total solids content of a 9.1/1 gelatin (275 bloom)/polyphosphate complex coacervation system affects the coacervation reaction. Each of the three continous curves shown encloses the region in which three systems of varying initial total solids content experience complex coacervation. The line that divides each curve represents the total solids content of the system before coacervation. Points above these lines are total solids content of the coacervate; points below these lines are total solids content of the equilibrium liquid. Note how the continuous curve that defines the coacervation region for a specific system expands as the initial total solids content decreases. This reflects the increase in coacervation pH range and coacervate solids content caused by dilution of the coacervation system. When the initial total solids content was reduced to 1.22 and 2.44 percent, some coacervates isolated had such high solids content that they behaved more like solids than liquids.

The gelatin (275 bloom) - polyphosphate system described by Figures 1 and 2 can concentrate in the coacervate phase a high percentage of the total solids in the system. Figure 5 shows that this system at pH 3.7 has 74 percent of its polyphosphate in the coacervate; at pH 3.8, 85 percent of the gelatin is there. These are maximum values that decrease rapidly as the pH is raised or lowered from 3.7 - 3.8. Similar curves are obtained with all coacervate systems that have an initial gelatin (275 bloom)/Polyphosphate ratio of 9.1/1 (w/w). Coacervates isolated from such systems have a higher gelatin/polyphosphate ratio than the equilibrium liquid at all coacervation pH's. Furthermore, the gelatin/-polyphosphate ratio in the coacervate phase increases significantly as the pH of coacervation rises above pH 4.0.

In addition to pH and initial total solids content, the gelatin/polyphosphate ratio of a coacervation system is an important parameter that affects the coacervation reaction. This was demonstrated by using a series of samples with fixed total solids content (4.88 percent), but varying gelatin (275 bloom)/Polyphosphate (w/w) ratios: 0.43/1, 1/1, 3/1, 9.1/1, and 18/1. The first two ratios gave no coacervate at pH 3.3 or 4.4, and were not examined further. It is possible that they would have formed a coacervate if diluted. The other three ratios gave a complex coacervate. However, Figure 6 shows that the continuous curves of total solids content versus pH for these three systems differ significantly. The curves for gelatin (275 bloom)/phosphate ratios of 18/1 and 3/1 fall completely within the curve for the 9.1/1 ratio. Thus, the 9.1/1 gelatin/polyphosphate system forms a more concentrated coacervate over a wider pH range than the other two ratios. Table II lists the maximum θ value for each of these three systems and the pH at which it falls. The 9.1/1 gelatin (275 bloom) polyphosphate system clearly has a much higher maximum coacervation intensity than the other two systems. The pH at which maximum coacervation intensity occurs increases with increasing gelatin/polyphosphate ratio.

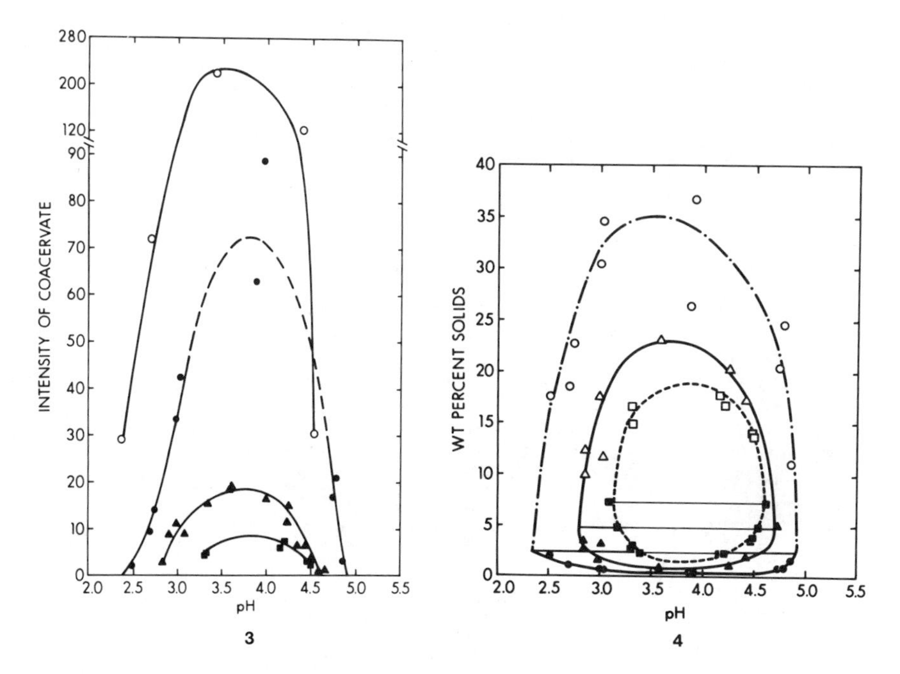

Figure 3. Plots of coacervation intensity versus pH at 55°C for four coacervation systems with a constant 9.1/1 (w/w) gelatin (275 bloom)/ polyphosphate ratio, but varying initial total solids content: ■, 7.32 percent initial total solids; ▲, 4.88 percent initial total solids, ●, 2.44 percent initial total solids; O, 1.22 percent initial total solids.

Figure 4. Plots of total solids content versus pH at 55°C for three 9.1/1 (w/w) gelatin (275 bloom)/polyphosphate coacervation systems. Open points: total solids content of coacervate; closed points: total solids of equilibrium liquid. Initial total solids content: □, ■, 7.32 percent; Δ, ▲, 4.88 percent; O, ●, 2.44 percent.

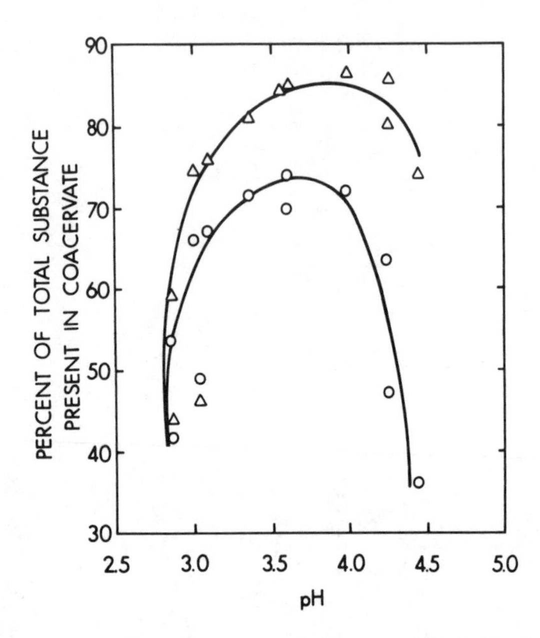

Figure 5. Plot of the fraction of gelatin (275 bloom), Δ , and polyphosphate, O, located in coacervates isolated at 55°C from a 4.4 percent gelatin and 0.48 percent polyphosphate coacervation system.

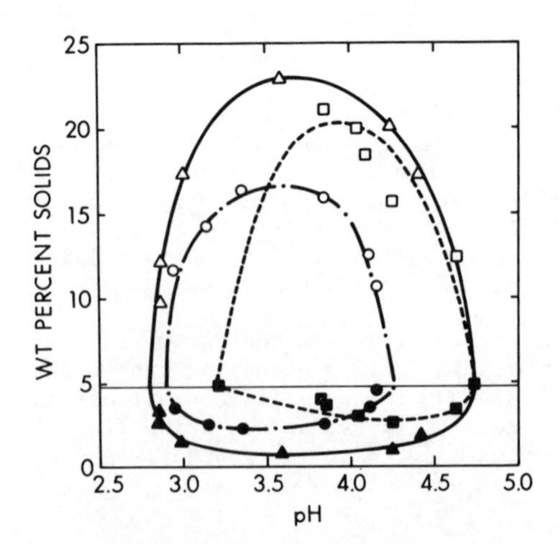

Figure 6. Plots of total solids content versus pH at 55°C for three coacervation systems with varying gelatin (275 bloom)/-polyphosphate ratios (w/w), but constant total initial solids of 4.88 percent. Open points: total solids content of coacervate; closed points: total solids of equilibrium liquid: O, ●, 3/1 ratio; Δ, ▲, 9/1 ratio; □, ■, 18/1 ratio.

Table II. Tabulation of Maximum Coacervation Intensity (θ) Values for Coacervation Systems with Varying Gelatin (275 Bloom)/Polyphosphate Ratios*

Gelatin/Polyphosphate Ratio, w/w	Maximum θ Value	pH for Maximum θ
3/1	4.6	3.3 - 3.6
9.1/1	19	3.5 - 3.7
18/1	3.5	4.1 - 4.3

*Total solids (gelatin plus polyphosphate) fixed at 4.8 wt. percent. pH adjusted with acetic acid.

Another factor that affects the gelatin/polyphosphate complex coacervation reaction is gelatin bloom strength. All data presented up to this point were obtained with a 275 bloom strength gelatin sample. In order to determine how bloom strength affects the coacervation process, experiments were made with 150, 275 and 300 bloom gelatins. The solids content in all cases was fixed at 4.4 percent gelatin and 0.48 percent polyphosphate. Figure 7 is a plot of total solids content versus pH for the coacervate and equilibrium liquid phases from each of the three bloom strength gelatins. The line at 4.88 percent total solids divides the curves into two parts. Points falling above this line are total solids contents of the coacervate; points below this line are total solids contents of the equilibrium liquid. The three continuous curves show appreciable overlaps although some differences are apparent. For example, the coacervation range for the system containing 300 bloom gelatin extends about 0.2 pH unit lower than that of the 150 bloom gelatin. The 150 bloom gelatin sample also gives an equilibrium liquid with somewhat more total solids than the two higher bloom gelatins. Although these differences appear to be relatively small, plots of coacervation intensity, θ, versus pH show more clearly that bloom strength has a significant effect on the gelatin-polyphosphate coacervation reaction. Figure 8 contains these plots. The 300 bloom gelatin sample has a maximum θ nearly three times greater than that of the 150 bloom gelatin sample. This indicates that higher bloom gelatins experience a much stronger complex coacervation reaction with polyphosphate than 150 bloom gelatin. This difference was not apparent in the total solids plot of Figure 7, because this plot did not disclose that the volume of coacervate formed by the 150 bloom gelatin was significantly less than that formed by the 275 or 300 bloom gelatins under the same experimental conditions. The 150 bloom gelatin gives a coacervate with essentially the same total solids content as that formed by higher bloom strength gelatins, but there is less of it. Thus, the equilibrium liquid for this system contains more gelatin than the equilibrium liquid isolated from coacervation systems involving higher bloom strength gelatins. This appears as a slight concentration increase in Figure 7, because the large volume of the equilibrium phase relative to the coacervate phase masks the difference.

As with the other coacervate systems examined in this study, the different bloom strength gelatins gave coacervates with gelatin/polyphosphate ratios greater than that in the equilibrium liquid

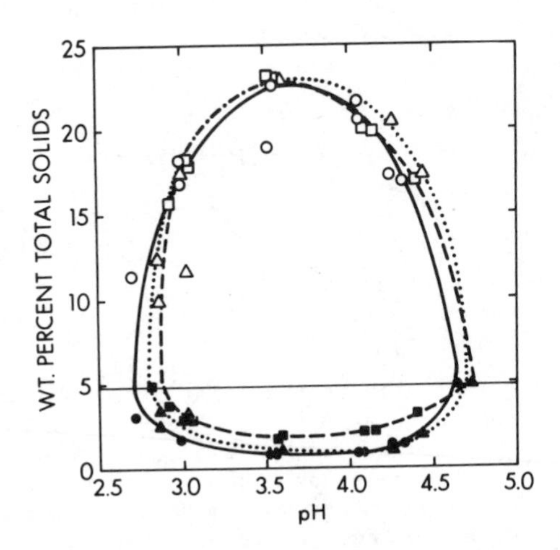

Figure 7. Plots of total solids content versus pH at 55°C for three coacervation systems with constant 4.4 percent gelatin and 0.48 percent polyphosphate content, but varying gelatin bloom strength. Open points: total solids content of coacervate; closed points: total solids content of equilibrium liquid: O, ●, 300 bloom; □, ■, 275 bloom; △, ▲, 150 bloom.

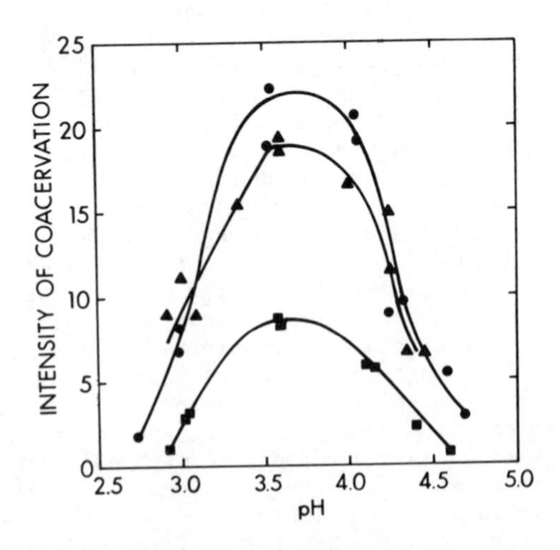

Figure 8. Plots of coacervation intensity versus pH for 4.4 percent gelatin and 0.48 percent polyphosphate mixtures prepared from three different gelatin samples: ■, 150 bloom gelatin; ▲, 275 bloom gelatin; ●, 300 bloom gelatin.

at all coacervation pH values. This ratio in the coacervate phase increases as the coacervation pH increases, especially above pH 4.0.

Discussion

The material balance and coacervation intensity data presented above demonstrate that the gelatin-polyphosphate mixtures used in this study act as classical complex coacervation systems. In some ways, such mixtures resemble gelatin-gum arabic mixtures. The pH range over which both types of systems form a coacervate is similar and varies with polyelectrolyte ratio. Dilution favors both coacervation reactions. Maximum coacervation intensity occurs at pH 3.5 to 3.7 for a 9.1/1 (w/w) gelatin/polyphosphate ratio, about the same pH at which a 1/1 (w/w) gelatin/gum arabic mixture appears to experience maximum coacervation (1). In spite of these similarities, gelatin/polyphosphate coacervates are distinctly different from gelatin-gum arabic coacervates. Gelatin/polyphosphate coacervates generally have a higher total solids content. Dilute mixtures may give coacervates that are concentrated enough to act more like solids than liquids. It is not surprising that polyphosphates reportedly precipitate proteins (6).

Another difference is that the solids content of gelatin/-polyphosphate coacervates is predominately gelatin whereas the solids content of gelatin/gum arabic coacervates is essentially a 1/1 (w/w) mixture of gelatin and gum arabic. This difference in composition arises from differences in the nature of gum arabic and polyphosphate. Gum arabic is a high molecular weight organic polymer with an ionic equivalent weight of approximately 1,200 (1). This ionic equivalent weight varies with pH, since the degree of dissociation of the carboxyl groups along the gum arabic chain varies with pH. However, it always is large. In contrast, the polyphosphate used in this study is a relatively low molecular weight inorganic material with an ionic equivalent weight of 102, assuming complete dissociation. The low ionic equivalent weight of the polyphosphate has a marked effect on complex coacervate composition. Gelatin/gum arabic coacervates contain an essentially 1/1 (w/w) ratio of gelatin and gum arabic because gelatin and gum arabic have essentially the same ionic equivalent weight. Because polyphosphate has a low ionic equivalent weight, relatively little is needed to interact ionically with gelatin. Accordingly, the solids in such coacervates are predominately gelatin and should possess physical properties that closely approach those of gelatin. The solids isolated by drying a gelatin/gum arabic coacervate should have physical properties characteristic of a 50/50 mixture of gelatin and gum arabic. These properties should differ from those of the solids isolated from a gelatin/polyphosphate coacervate. Of course, ionic interactions in a gelatin/polyphosphate or gelatin/-gum arabic coacervate may affect the physical properties of the solids isolated from both types of coacervates, but this remains to be defined.

Coacervates isolated from gelatin/polyphosphate coacervation systems normally have gelatin/polyphosphate ratios (w/w) between 10/1 and 12/1. This approaches the ratio of gelatin/polyphosphate equivalent weights and reflects the need for ionic equivalence in complex coacervates. A 9.1/1 gelatin/polyphosphate mixture gives a more intense coacervation reaction than 3/1 or 18/1 mixtures (Figure 6), because it more closely approaches the exact ionic equivalence ratio. Mixtures with ratios between 10/1 and 12/1 should give even more intense coacervation reactions, since they fall closer to the exact ionic equivalence ratio. This ratio shifts somewhat with pH, especially above pH 4.0, presumably due to changes in degree of ionization of gelatin and polyphosphate with pH.

The high solids content of many gelatin/polyphosphate coacervates is due to the high coacervation intensity experienced by such mixtures. This is another result of the low ionic equivalent weight of polyphosphate. Coacervation intensities found in this study consistently fell above 10 and sometimes exceeded 200. The few data that are in the literature indicate gelatin/polyphosphate mixtures interact more intensely than either gelatin/gelatin (5, 7) or gelatin/gum arabic (1) mixtures. Gelatin/gelatin coacervation systems generally have θ values well below 10 (5), and experience a relatively weak coacervation interaction. Enrichment data reported for a gelatin/gum arabic coacervation system (1) indicate that such systems interact more intensely than a gelatin/gelatin system, but less intensely than a gelatin/polyphosphate system.

Literature Cited

1. Bungenberg de Jong, H. G., in "Colloid Science", Vol. II, H. R. Kruyt, ed., Elsevier Publishing Co., N. Y., 1949, Chap. X.

2. Green, B. K. and Schleicher, L., U.S. Patent 2,800,457, July 23, 1957.

3. Horger, G., U.S. Patent 3,872,024, March 18, 1975.

4. Chen, P. S., Toribara, T. Y., and Warner, H., Anal. Chem. 28, 1756 (1956).

5. Veis, A., and Aranyi, C., J. Phys. Chem., 64, 1203 (1960).

6. Van Wazer, J. R., "Phosphorous and Its Compounds", Vol. I, Interscience Publishers, N.Y., 1958, pg. 466.

7. Veis, A., Bodor, E., and Mussell, S., Biopolymers 5, 37-59 (1967).

RECEIVED January 24, 1986

Complex Coacervate Formation Between Acid- and Alkaline-Processed Gelatins

D. J. Burgess[1] and J. E. Carless

The School of Pharmacy, University of London, 29-39 Brunswick Square, London WC1 1AX, United Kingdom

Microeletrophoretic mobility profiles of an acid and an alkaline processed gelatin were used to determine the optimum pH and ionic strength requirements for complex coacervation of these gelatins. The pH range where the gelatins carried opposite charges and therefore may be capable of forming coacervates was narrow (pH 4.8 to 8.3). Within this pH range the charge carried by the gelatins was very low, unless the ionic strength of the medium was below 1mM. Coacervation was therefore expected to occur in the above pH range at low ionic strength. However, coacervation was not detected in mixtures of the gelatins, regardless of the pH and ionic strength conditions, unless the temperature was reduced below the gelation temperature of the gelatins. Evidence obtained from photon correlation spectroscopy measurements suggested that soluble aggregrates may form between the gelatin mixtures at temperatures above their gelation temperature. This is in accordance with the 'dilute phase aggregrate' model of coacervation, Veis and Aranyi (1).

Oppositely charged polyions in aqueous media may interact spontaneously to form complex coacervates, Bungenberg de Jong (2). Bungenberg de Jong explained complex coacervation on the basis of the random coil macromolecule. He considered that coacervation occurred as a result of electrostatic interaction forces, ignoring the polymeric nature of the polyions involved. This was an important omission however, as small ion pairs such as sodium chloride dissociate in water due to the large hydration shell possible. Diamond (3) proposed that 'water structure-enforced' ion pairing would result on mixing polycations and polyanions, involving both electrostatic and hydrophobic interactions.

A theoretical treatment of complex coacervation was worked out by Voorn (4,5) and Overbeek and Voorn (6). They explained the process as a competition between the electrical attractive forces tending to accumulate the charged polyions and entropy effects

[1]Current address: Department of Pharmacy, University of Nottingham, University Park, Nottingham NG7 2RD, United Kingdom

0097-6156/86/0302-0251$06.00/0

tending to disperse them. It has been shown (6) using the Debye-Huckel equations (7,8) for the electrical interaction term and the Flory-Huggins theory (9-13) for the entropy term that the critical conditions for complex coacervation were obtained when $\sigma^3 r < 0.5$. That is, when either the charge density (σ) or the molecular weight (r) or both were sufficiently large. Their theory was based on the following assumptions: that there was a random chain distribution of macromolecules in both phases; that solvent-solute interactions were negligible, i.e. the Huggins interaction parameter was negligible; and that the interactive forces were of a distributive nature with the system behaving as though the charges were free to move. The macromolecular skeins of two oppositely charged polyions associate together as a result of the electrostatic forces to form a coacervate phase, and while doing so entrap water between their loops, known as occlusion water. As a consequence of this entrapped water the coacervate is liquid in nature and the process is readily reversible.

Veis and Aranyi (1) however, have reported complex coacervation in solutions of oppositely charged gelatins where the above requirements were not met. In order to explain this effect they proposed that in their system, complex coacervation occurred by a different mechanism i.e. spontaneous aggregation of the oppositely charged gelatins took place by electrostatic interaction upon mixing, to form aggregates of low configurational entropy; and these aggregates then rearranged to form a coacervate phase. This second reaction occurs more slowly and is driven by the gain in configurational entropy which results in the formation of a randomly mixed concentrated coacervate phase and the dilution of the non-aggregate phase. Thus the electrostatic free energy change and the entropy change play opposite roles to those in the Voorn-Overbeek theory.

Veis (14) supported his theory with light scattering data which indicated the presence of aggregates in the equilibrium fluid after phase separation of mixtures of isoionic gelatins. Kurskaya *et al* (15) have also reported turbidimetric evidence for the presence of what they term 'soluble complexes' in solutions of oppositely charged isoionic gelatins. However, it is well known that individual salt-free gelatin solutions form aggregates at their isoionic pH due to intermolecular attractive forces resulting from charge fluctuations (16).

The 'dilute phase aggregate' model has been used to describe the complex coacervation of gelatin and sodium alginate (17). Tomlinson and Davis (18-20) and Wilson *et al* (21) proposed that ion pair formation between sodium cromoglycate and dodecylbenzyldimethylammonium chloride occurs prior to complex coacervation. They interpreted the mechanism as being ion association reinforced by a strong hydrophobic effect, according to the theory of Diamond (3). It has also been shown (22) that aggregates form in dilute solution between negatively and positively charged polyvinylalcohol. The mechanism of interaction was interpreted according to the Voorn-Overbeek approach but including the Huggings interaction parameter (23,24).

The aim of this study is to prepare complex coacervates between oppositely charged gelatins and to analyse the process for theoretical significance. In a previous publication (25) the effect of pH and ionic strength on the coacervate yield of gelatin and acacia mixtures was explained by the effect of these variables on the charge of the two polyions involved. A method was described for the prediction of complex coacervation using microelectrophoresis to measure the charge on the polyions. This study reports the use of this method to optimise conditions for complex coacervation between oppositely charged gelatins.

Materials and Methods

Two types of gelatin were obtained from Gelatin Products Ltd., U.K., Type A (acid processed) gelatin and Type B (alkali processed) gelatin. The gelatins had the following characteristics: Type A; Bloom No. 256, isoelectric pH 8.3, $M\bar{n}$. 4.7 x 10^4, and ash content 0.2% w/w.; Type B; Bloom No. 250, isoelectric pH 4.8, $M\bar{n}$. 4.6 x 10^4, and ash content 1.1% w/w. The isoelectric pH values were measured by microelectrophoresis and by ion exchange. The $M\bar{n}$. was measured by membrane osmometry using a Wescan Model 231 membrane osmometer, Wescan Instruments Inc.

The acacia used was of British Pharmacopoeial quality and had an ash content of 3.2% w/w and a sulphated ash value of 4.9% w/w. All gelatin and acacia solutions were deionised by mixing with Amberlite resins IRA-400 + IR-120 for 30 min at $40^{\circ}C$ prior to use, unless otherwise stated. This method was adapted from Janus et al (26). The potassium thiocyanate used was of analytical grade from Koch-Light Laboratories Ltd.

Microelectrophoresis. A Zeta-Meter (Zeta Meter Inc.) was used in conjunction with a Plexiglas cell. Microelectrophoresis was conducted at 1 mM NaCl unless otherwise stated. Constant ionic strength was maintained as the pH was varied using 1 mM NaOH and 1 mM HCl solutions. The polyions were adsorbed onto Minusil (colloidal silica), of particle size 2.7 μm (geometric weight-mean diameter, with a geometric standard deviation of 0.72 μm) prior to microelectrophoresis. A 0.02% w/v polyion solution and a 0.01 % w/v Minusil suspension were used. The electrophoretic mobility was the mean of at least 20 readings and the coefficient of variation was less than 5%.

Coacervate Yield Determination. Gelatin coacervate mixtures, formed as described in the results section, were centrifuged to separate the coacervate and equilibrium phases, using an MSE High Speed 18 centrifuge at a speed of 1000 to 2000 rpm, set at the appropriate final temperature, of the coacervate mixture. The equilibrium phase was decanted and the coacervate phase was then heated to $40^{\circ}C$ and deionised if necessary (as described above), so that only the weight of the polyions was determined and not that of any associated salt ions. The resultant solution was transferred into a weighed evaporating dish and dried to constant weight at $60^{\circ}C$.

Photon Correlation Spectroscopy (PCS). A Malvern model 4300 photon correlation spectrometer with a 64 channel type K7027 Loglin correlator was used in conjunction with a Liconix He/Cd laser operating at 441.6 nm. The temperature was controlled to $0.1^{\circ}C$. The data were processed to give values for the equivalent spherical hydrodynamic diameter and for the normalised variance of distribution (NVD). The NVD value gives an indication of the degree of polydispersity of the sample. All solutions were filtered through 0.45 μm Millipore filters into PCS cells which were immediately covered. The final solutions in the PCS cells were allowed thirty minutes to equilibrate at 40°C before any measurements were taken. Deionisation was carried out by mixing with Amberlite resins as before.

Results and Discussion

Microelectrophoresis of the Gelatins. The electrophoretic mobility profiles of the two gelatins are shown in Figure 1. The pH range where the gelatins bear opposite charges and therefore might be expected to form complex coacervates is pH 4.8 to pH 8.3. Over this pH range the electrophoretic mobility of Type A gelatin is positive but small, while the electrophoretic mobility of Type B gelatin decreases rapidly from zero to high negative values. At the electrical equivalence point (EEP) of the gelatins (pH 5.4, calculated from Figure 1) i.e. the pH where the gelatins carry equal but opposite charges; both gelatins have an electrophoretic mobility of $0.5 \times 10^{-8}\ m^2\ s^{-1}$. This relatively small charge may be insufficient to bring about coacervation (25). If complex coacervation does occur between the gelatins the pH range is expected to be limited due to the large imbalance in charge between the two gelatins at pH values distant from the EEP (see Figure 1).

The effect of ionic strength on the electrophoretic mobility of the gelatin is shown in Figure 2. The electrophoretic mobilities of both the gelatins increase significantly when the ionic strength is decreased below 1 mM. It may therefore be concluded that coacervation will occur more readily between the gelatins at very low ionic strength. This is in agreement with the results of Veis and co-workers (1,14,28) who used deionised gelatins to produce Type A/Type B gelatin coacervates under ion-free conditions. It is predicted that complex coacervation of a 1:1 Type A/Type B gelatin mixture will occur only at low ionic strength, within a pH range of approximately pH 5.0 to pH 6.0, with maximum coacervation occurring at pH 5.4.

Formation of Gelatin/Gelatin Complex Coacervates. Ion-free (1:1) gelatin/gelatin mixtures were prepared at pH 5.4 (the predicted optimum conditions for coacervation) over a total concentration range of 0.2 to 4.0% w/v at 40°C. However coacervation was not evident. The pH range, pH 4.8 to pH 8.3 and the ionic strength range, zero to 10 mM were next investigated. Again coacervation was not apparent. The temperature of the gelatin mixtures was reduced below 35°C (the gelation temperature of the gelatins) and complex coacervates formed within specific limits: pH (pH 5.2 to pH 5.8); ionic strength (0 to 3 mM); total gelatin concentration (0.2 to 2.% w/v); and polyion mixing ratio (0.75 to 1.50, Type A/Type B gelatin).

The effects of pH and ionic strength are in agreement with the predictions made using the microelectrophoresis data obtained on these gelatins. Since the charge on the gelatins is a limiting factor for coacervation it is conjectured that electrostatic interaction is a pre-requisite for the build up of these gelatin/gelatin coacervates, supporting the 'dilute phase aggregate' theory (1).

Interaction between the gelatin molecules at sufficiently high concentration may result in the build up of large gel-type networks in the equilibrium fluid, and therefore the gain in energy on coacervate phase formation will be reduced and coacervation will be suppressed (self-suppression). The maximum gelatin concentration tolerated was low, up to 2.0% w/v. This may be a consequence of the large extent of interaction possible between these very similar gelatin molecules.

The final temperature to which a mixture was reduced and the length of time allowed for equilibration at the final temperature both dramatically affected the coacervate yield. Figure 3 shows the effect of final temperature on the coacervate yield of 1:1 mixtures of the gelatins (1% w/v). All the mixtures were equilibrated at their final temperature for a period of six hours. The lower the final temperature the higher the coacervate yield obtained, to a maximum at around 15°C. Initially only the high molecular weight gelatins coacervate but as the temperature is reduced conditions become favourable for coacervation of the lower molecular weight species, thus explaining the increase in yield observed. Figure 4 shows the effect of the time allowed for coacervation to occur on the coacervate yield of 1:1 mixtures of the gelatins (1% w/v) held at a final temperature of 15°C. The longer the time allowed, up to approximately six hours, the higher the coacervate yield. Coacervation is normally considered to be a spontaneous process, however gelation forces which operate between gelatin molecules at low temperatures take time to develop. It appears that both electrostatic attraction and gelation forces are necessary for the formation of gelatin/gelatin coacervates.

It is postulated that this process is consistent with the 'dilute phase aggregate' model (1) forming Type A/Type B aggregates initially when mixed at 40°C, which rearrange to form a coacervate phase when the temperature is reduced below 35°C.

Photon Correlation Spectroscopy Studies. In order to investigate possible aggregate formation between the gelatins at 40°C photon correlation spectroscopy was used. The following pairs of Type A and Type B gelatin solutions (0.5% w/v) were prepared and studied by PCS both individually and as mixtures and the values obtained for the average particle diameter and the polydispersity were compared:

(i) Deionised solutions prepared at 40°C;
(ii) Deionised solutions heated to 50°C for thirty minutes, then cooled to 40°C;
(iii) Deionised solutions prepared at 40°C in the presence of 1 M potassium thiocyanate;
(iv) Non-deionised solutions prepared at 40°C.

These results are given in Table 1. Both deionised and non-deionised gelatin solutions were studied, to compare coacervating (deionised) and non-coacervating (non-deionised) mixtures. Type A and Type B gelatin contain a range of molecular weights as is evident from their large NVD values (Table 1). This polydispersity made interpretation of the results difficult. In solutions containing only one type of gelatin the molecules also tend to form into aggregates, especially around the isoelectric point of the gelatins (16) which also made interpretation of the results difficult. This type of aggregation is particularly apparent in Type A gelatin solutions as can be seen by the significant reduction in the average particle size when the solutions were treated to decrease aggregation, either by preheating or by the addition of

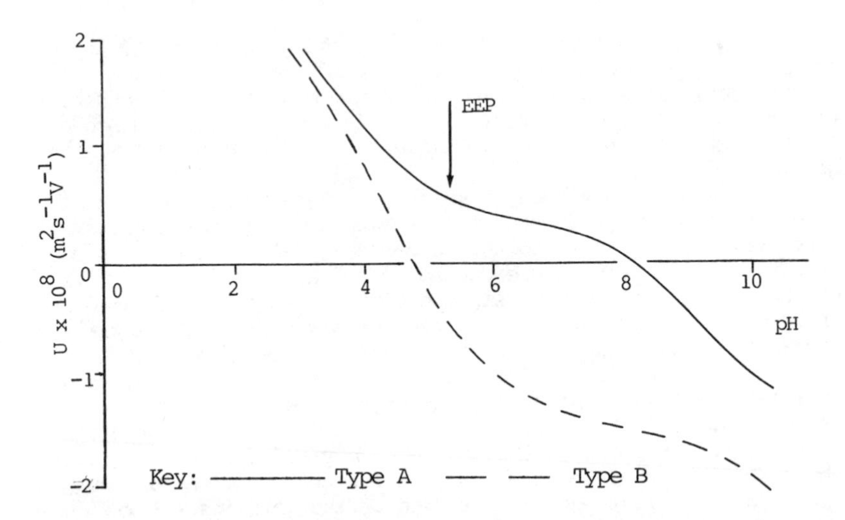

Figure 1. The effect of pH on the electrophoretic mobility of Types A and B gelatin.

EEP - The Electrical Equivalence Point

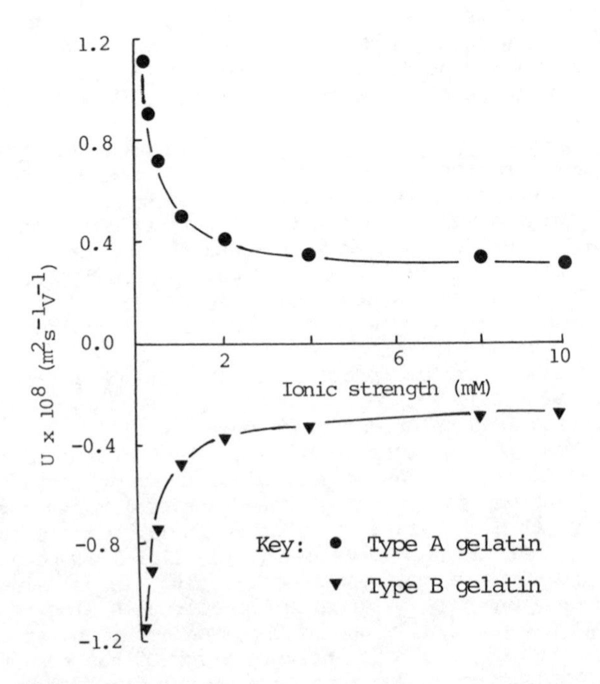

Figure 2. The effect of ionic strength on the electrophoretic mobility of Types A and B gelatin.

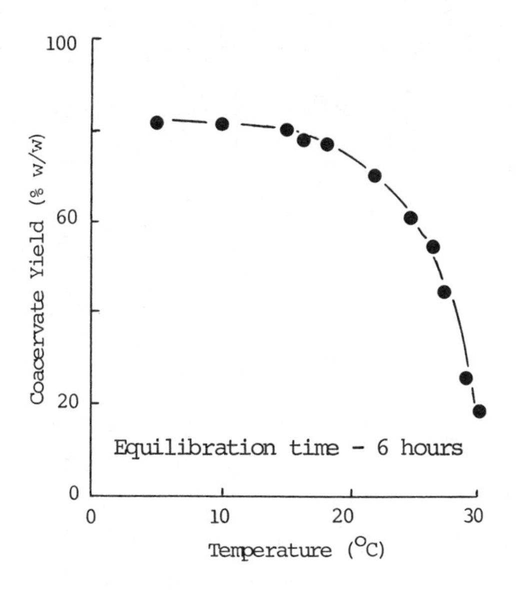

Figure 3. The effect of final temperature on gelatin/gelatin coacervate yield.

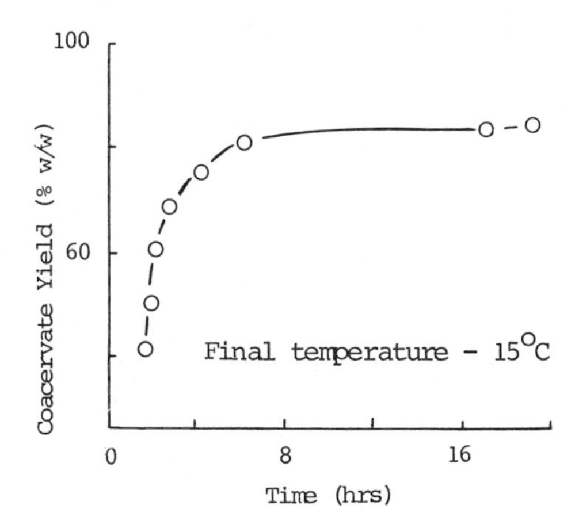

Figure 4. The effect of time allowed for coacervation to occur on gelatin/gelatin coacervate yield.

TABLE I

Particle Diameter and Polydispersity of 0.5% w/v Solutions of Type A and Type B Gelatin, Before and After Mixing at 40°C (Equal Concentrations)

Treatment of Gelatin Solutions		Particle Diameter (nm)			Polydispersity (N.V.D.)		
		Type A Gelatin	Type B Gelatin	Mixture	Type A Gelatin	Type B Gelatin	Mixture
(i)	Deionised	217 ± 2	58.8±0.5	62.9±0.5	0.66±0.09	0.34±0.08	0.22±0.03
(ii)	Deionised and pre-heated at 50°C	50.1±1	46.1±0.5	60.5±0.5	0.41±0.016	0.39±0.07	0.22±0.02
(iii)	Deionised + 0.1% w/v KCNS	73.7±2.5	64.6±3.5	68.2±2.5	0.73±0.10	0.67±0.10	0.78±0.10
(iv)	Non-deionised	438 ± 20	57.9±2	197 ±20	0.77±0.10	0.41±0.10	0.93±0.12

potassium thiocyanate (see Table 1). The average particle size of Type B gelatin was also reduced by preheating. The presence of potassium thiocyanate increased the average particle size of Type B gelatin, this is probably a consequence of expansion of the gelatin molecules in the presence of this salt which is known to prevent inter- and intra-molecular interactions in gelatin solutions.

It is difficult to draw any conclusions from a study of the average particle sizes of the gelatin solutions and mixtures. However the changes in polydispersity which occurred on mixing pairs of gelatin solutions indicated in some cases that interaction had occurred. The mixtures of the deionised gelatins had NVD values which were smaller than those of the individual gelatin solutions, suggesting that interaction may have occurred. In the absence of interaction an increase in polydispersity would be expected to occur on mixing two macromolecular solutions of different average particle size. In the gelatin mixtures which contained salt ions and as a result were unable to undergo coacervation on temperature reduction, interaction was not apparent. These mixtures were more polydispersed than the individual solutions.

The changes in polydispersity observed on mixing the deionised gelatin solutions may be due to the formation of aggregates by electrostatic interaction between the oppositely charged gelatins, supporting the 'dilute phase aggregate' model (1). Since the two gelatins are similar in structure, strong bonding may occur between the oppositely charged molecules giving the 'aggregates' compact structures. The small average particle sizes obtained for these 'aggregates' suggests that they are composed of only a few molecules and probably only contain one pair of molecules per aggregate. This is in agreement with the results of Veis (14) and those of Kurskaya et al (15) which indicated that on mixing equal amounts of electrically equivalent gelatins under ion-free conditions ion pairs formed.

Conclusions

Microelectrophoresis data were successfully used to predict the effect of pH and ionic strength on gelatin/gelatin coacervation. This coacervate system was also shown to be strongly dependent on temperature, temperature reduction being necessary to initiate coacervation. This is probably a consequence of the charges carried by the gelatins, which are too low to bring about phase separation by electrostatic interaction alone. Measurements made by photon correlation spectroscopy on gelatin/gelatin mixtures at 40°C suggest that aggregates form between the oppositely charged gelatins on mixing, although phase separation was not evident until the temperature was reduced below the gelation temperature of the gelatins.

The gelatin/gelatin coacervate systems studied appear to fit the Veis-Aranyi 'dilute phase aggregate' model (1), forming aggregates by electrostatic interaction which may then associate to form a separate coacervate phase by the aid of gelation forces on temperature reduction.

Acknowledgments

The authors thank: Dr. T.L. Whateley, Pharmacy Department, University of Strathclyde, Glasgow, U.K. for his advice and helpful discussions regarding the photon correlation spectroscopy studies; Gelatin products Ltd. for samples of gelatins; and the Nicholas Drug Consortium for financial assistance.

Literature Cited

1. Veis, A.; Aranyi, C.J. J. Phys. Chem. 1960, 64, 1203.
2. Bungenberg de Jong, H. G. In "Colloid Science"; Kruyt, H.R., Ed.; Elsvier: Amsterdam, 1949; Vol II, Chapters VIII & X.
3. Diamond, R. M. J. Phys. Chem. 1963, 67, 2513.
4. Voorn, M. J. Rec. Trav. Chim. 1956, 75, p. 317, 405, 427, 925, & 1021.
5. Voorn, M. J. Fortschritte Hochpolm. Forschy Bd. 1959, 1.S., 192.
6. Overbeek, J. TH. G.; Voorn, M. J. J. Cellular Compar. Physiol. 1957, 49, Suppl. 1, 7.
7. Debye, P.; Huckel, E. Physik. Z. 1923, 24, 185.
8. Debye, P.; Huckel, E. Physik. Z. 1924, 25, 49.
9. Flory, P.J. J. Chem. Phys. 1941, 9, 660.
10. Flory, P. J. J. Chem. Phys. 1942, 10, 51.
11. Flory, P. J. J. Chem. Phys. 1944, 12, 425.
12. Huggins, M. L. J. Phys. Chem. 1942, 46, 151.
13. Huggins, M. L. J. Am. Chem. Soc. 1942, 64, 1712.
14. Veis, A. J. Phys. Chem. 1963, 67, 1960.
15. Kurskaya, E. A.; Vainerman, E. S.; Timofeeva, G. I.; Rogozhin, S. U. J. Colloid Polymer Sci. 1980, 258, 1086.
16. Boedtker, H; Doty, P. J. Phys. Chem. 1954, 58, 968.
17. Wajnerman, E. S.; Grinberg, W. Ja.; Tolstogusow, W. B. Kolloid-Z. u. Z. Polymere 1972, 250, 945.
18. Tomlinson, E.; Davis, S. S. J. Colloid Interface Sci. 1978, 66, 335.
19. Tomlinson, E.; Davis, S. S. J. Colloid Interface Sci. 1980, 74, 349.
20. Tomlinson, E.; Davis, S. S. J. Colloid Interface Sci. 1980, 76, 563.
21. Wilson, C. G.; Tomlinson, E.; Davis, S. S.; Olejnik, O. J. Pharm. Pharmacol. 1981, 31, 749.
22. Sato, H.; Nakajima, A. Colloid & Polmer Sci. 1974, 252, 944.
23. Nakajima, A.; Sato, H. Biopolymers 1972, 11, 1345.
24. Sato, H.; Nakajima, A. Colloid & Polmer Sci. 1974, 252, 294.
25. Burgess, D. J.; Carless, J. E. J. Colloid Interface Sci. 1984, 98, 1.
26. Janus, J. W.; Kenchington, A. W.; Ward, A. G. Research (Lond.) 1951, 4, 247.
27. Veis, A. J. Phys. Chem. 1961, 65, 1798.
28. Veis, A.; Bodor, E.; Mussell, S. Biopolymers 1967, 5, 37.

RECEIVED July 8, 1985

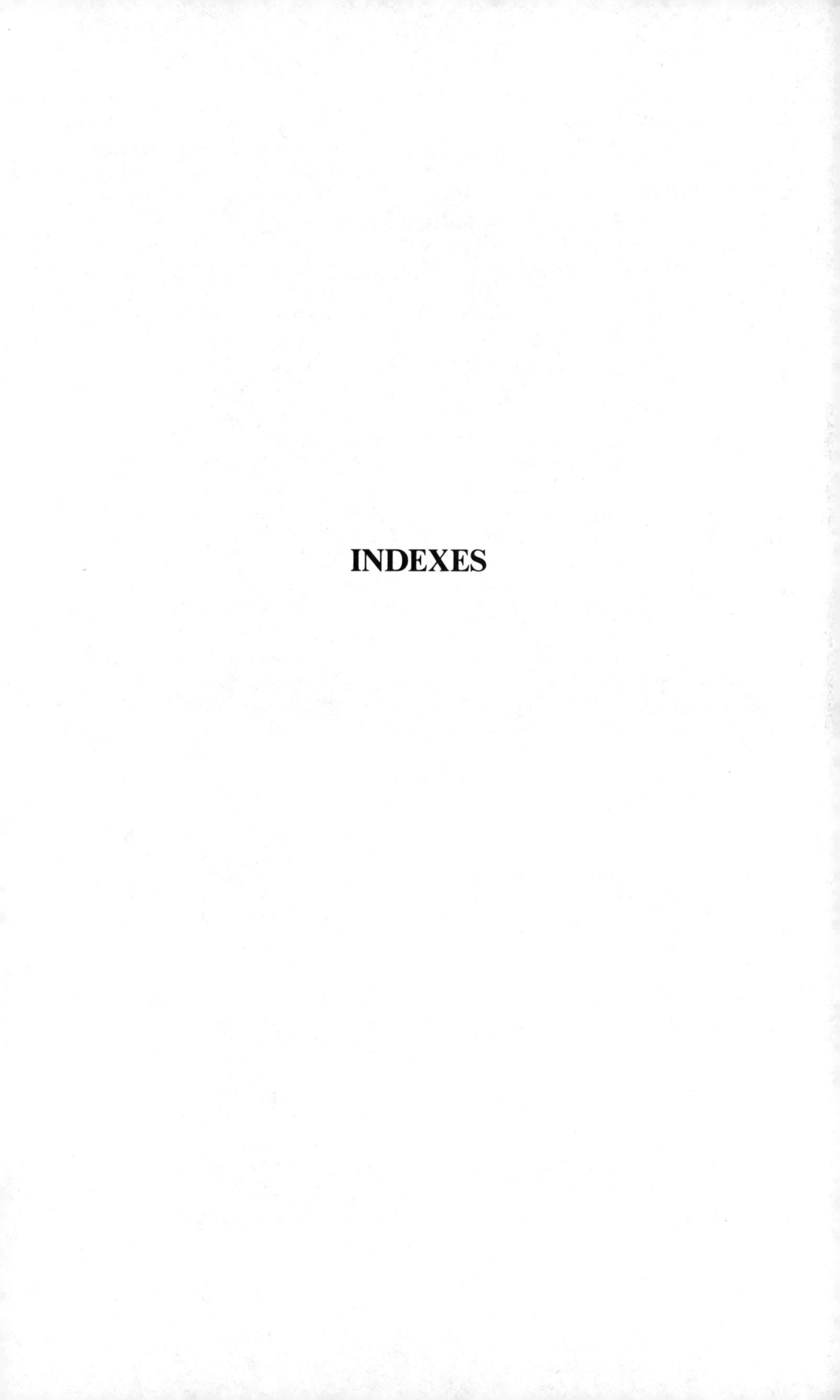

INDEXES

Author Index

Subject Index

M

N

O

P

Production by Anne Riesberg
Indexing by Karen McCeney
Jacket design by Pamela Lewis

Elements typeset by Hot Type Ltd., Washington, DC
Printed and bound by Maple Press Co., York, PA